REACTIVATING ELEMENTS

ELEMENTS   *A series edited by Stacy Alaimo and Nicole Starosielski*

REACTIVATING ELEMENTS

CHEMISTRY, ECOLOGY, PRACTICE

EDITED BY
DIMITRIS PAPADOPOULOS,
MARIA PUIG DE LA BELLACASA,
& NATASHA MYERS

DUKE UNIVERSITY PRESS   Durham and London   2021

© 2021 Duke University Press. All rights reserved
Printed and bound by CPI Group (UK) Ltd, Croydon, CR0 4YY
Typeset in Chaparral Pro and Knockout by BW&A Books, Inc.

Library of Congress Cataloging-in-Publication Data
Names: Papadopoulos, Dimitris, [date] editor. | Puig de la Bellacasa,
María, editor. | Myers, Natasha, [date] editor.
Title: Reactivating elements : chemistry, ecology, practice / edited by
Dimitris Papadopoulos, María Puig de la Bellacasa, and Natasha Myers.
Other titles: Elements (Duke University Press)
Description: Durham : Duke University Press, 2021. |
Series: Elements | Includes bibliographical references and index.
Identifiers: LCCN 2021014746 (print)
LCCN 2021014747 (ebook)
ISBN 9781478013440 (hardcover)
ISBN 9781478014362 (paperback)
ISBN 9781478021674 (ebook)
Subjects: LCSH: Chemical ecology. | Ecology. | Ecocriticism.
| Environmental chemistry. | BISAC: SCIENCE / Environmental
Science (see also Chemistry / Environmental) | SOCIAL SCIENCE /
Media Studies
Classification: LCC QH541.15.C44 R433 2021 (print) | LCC QH541.15.C44
(ebook) |
DDC 577/.14—dc23
LC record available at https://lccn.loc.gov/2021014746
LC ebook record available at https://lccn.loc.gov/2021014747

Cover art: "Agents of elemental rearrangement. At the edge of an urban oak
savannah, Dish with One Spoon Lands." Kinesthetic image by Natasha Myers,
2018, from Becoming Sensor: Ungrid-able Ecologies for a Planthroposcene,
www.becomingsensor.com.

Dimitris Papadopoulos would like to gratefully acknowledge the support of The Leverhulme Trust, UK (grant number RF-2018-338\4), as well as the Biotechnology and Biological Sciences Research Council (BBSRC), UK (grant number BB/L013940/1), and the Engineering and Physical Sciences Research Council (EPSRC), UK (grant number BB/L013940/1). María Puig de la Bellacasa would like to acknowledge the support of the Arts and Humanities Research Council (AHRC), UK (grant number AH/T00665X/1). The editors and publisher would like to thank the Faculty of Liberal Arts and Professional Studies, York University, Toronto, Canada, for the financial support it provided to this work. We are also grateful to the Centre for Interdisciplinary Methodologies (CIM), University of Warwick, UK, and the Institute for Science and Societies (ISS), University of Nottingham, UK, for their support.

FROM COSMOLOGY TO EPISTEME AND BACK

Dimitris Papadopoulos,
María Puig de la Bellacasa,
and Natasha Myers

Reactivating. Calling a recognized entity into a new situation; catalyzing new modes of thought and action; waking up new insight from the slumber of the familiar and the mundane. Reactivation, in the sense of reactivity, also recalls the agency of chemical substances. *Elements.* From the chemical elements to the elementals earth, air, fire, water—and back. Simultaneously evoking and blurring the ancient and modern, the elements have been brought back in recent years to speak in old and new tongues, in diverse contexts and practices, generating other ways of storying long-standing narratives. *Reactivating Elements* draws the nonlinear historical significance of elemental thought into contemporary practice, while inviting new provocations and curiosities about these ever-present and simultaneously elusive phenomena.

The elements are activated in the chapters of this collection in different ways, reflecting on and responding to the messy mixtures of elemental matters, energies, and modes of thought and practice shaping Earth's troubled ecologies today. These accounts engage the elements capaciously, taking up both the periodic table of the elements inscribed by the discipline of chemistry (Stengers) and those irreducible "elementary forms" that ground theories of social and political life (Helmreich, Hayden). They examine the elements as both the condition of existence and as what we dwell in, considering how the elements atmospherically mediate collective forms of belonging and social action, inciting new modes of "conspiracy" (Choy). By letting these multiple approaches to the elements react in this interpellating present, we call on their substantial and figurative potential to propel new ways of thinking, making, and doing that can respond to this planet's current predicaments.

Elemental thought is persistent. See how it bends time by looping ancient cosmologies into contemporary practices. Cosmologies, like Empedocles's four elements of earth, air, water, and fire, which parsed the elements into forces of nature and states of matter, remain salient today, informing Western ontologies and epistemologies as well as popular culture, while also grounding modern chemistry. Or consider how chemistry's periodic table of the elements remains ever-expanding, enabling the analysis and manipulation of matter at the scale of atoms, particles, and waves (Masco). The periodic table features prominently as one of chemistry's most precious devices, consolidating its disciplinary identity as a science and shaping the social and material worlds that we find ourselves in today. And at the same time, other elemental systems remain at work today in practice and culture. Chinese systems, with a world made of earth, fire, water, wood, and metal, continue to ground healing arts and philosophies in and well beyond China. Ancient alchemists with their passionate efforts to transmute the elements continue to spark speculative imaginaries today, while shamans and witches the world over continue to conjure other futures by allying with elemental forces (Puig de la Bellacasa). And while some elemental systems identify the elements as "natural" forces and substances external to the realm of human action, multiple Indigenous cosmologies, including those which ground practice and knowledge in communities today, approach the elements differently. Witness Indigenous land protectors on the front lines of local and global sovereignty movements who attest to the inseparability of land and body (Murphy). Or the affirmations of many Indigenous peoples of Turtle Island (so-called North America) who teach that water is not just a state of matter but rather that "water is life" (e.g., McGregor 2014). Perhaps what "moderns" chide as the "animisms" of non-Western ontologies stem precisely from Indigenous refusals to deanimate or disenchant the elements or to render them separate from human praxis and culture. Where some elemental cosmologies disavow such entanglements of nature and culture, Indigenous cosmologies (e.g., Kimmerer 2015; Simpson 2017) teach the deeper significance of what those of us caught within the constraints of colonial grammars gesture at when we say "naturecultures" (Haraway 1997).

As feminist science studies scholars have long affirmed, ontologies and epistemologies are fundamentally intertwined (see Barad 2007; Haraway 1988), such that the substances, forces, and energies that come to matter in a given world are inextricable from concepts, theories, and techniques engaged to know that world. Elements and elementary forms oscillate between substance and semiosis, figure and ground, episteme and cosmology, shaping how and what we come to know. It is in this sense that different epistemic iterations of the elements can be seen to activate diverging materialities, congeal distinct substances and meanings, and matter distinct worlds. Elemental thinking demands challenging entrenched epistemological and ontologies binaries—such as those that neatly distinguish what we see as living and nonliving (Schrader). Holding the chemical and the cosmological together may arouse the elements' transformative figurative potential. Many of the contributions in this collection document the different ways that elemental practices converge and diverge. And what these thinkers show is that however well-defined the elements appear to be, they are simultaneously elusive and unpredictable (Papadopoulos). If we let the elements lure our attention and alter our methods (Dumit), they can open up unexpected and generative insights into the actualities of our social, political, and ecological conditions and inspire interventions that activate other modes of world making (Bresnihan).

There is another reason why this collection explores multiple registers of the elements in one breath. The editors and the contributors came to meet around this project through intellectual, political, and personal friendships developed within our work in critical and feminist science and technology studies (STS). It is this community of practice that grounds our creative and undisciplined reactivation of the elements—the medium, we could say, that brought us together in the first place.[1] From a science and technology studies perspective, to call in the elements is to shape what comes to matter in technoscience: we call for a shift of attention to the techniques, processes, affects, and intensities that churn soils, airs, waters, and fires up together with organic and synthetic chemicals in order to attend more fully to bodily potentials, relations, toxicities, and harms. Yet the works included here also inherit and contribute to multilayered stories that go beyond an intervention in one research field. While rooted in perspectives, debates, methods, questions, and ways of thinking from within STS, *Reactivating Elements* connects and draws inspiration from a range of interconnecting discourses and bodies of thought, in particular the anthropology of science and technology (see, e.g., Choy 2011; Dumit 2004; Hayden 2003; Helmreich 2009; Masco 2004), political ecology (see, e.g., Bennett 2010; Ernstson and Swyngedouw 2019; Robbins 2012), environmental humanities (see, e.g., Alaimo 2016; Coole and Frost 2010; Rose 2004; Tsing 2015), more-than-human geography (see, e.g., Braun and Whatmore 2010; Clark 2011), history and philosophy of science (see, e.g., Barad 2007; Bensaude-Vincent and Stengers 1996; Lefèvre and Klein 2007), media studies (see, e.g., Parikka 2015; Peters 2015; Starosielski 2015), and, of course, the field of thought that brought back the elements into intellectual cultural imaginaries by developing contemporary elemental theory (see, e.g., Cohen and Duckert 2015; Macauley 2010; Neale, Phan, and Addison 2019).

What makes this collection distinctive in its contribution to current elemental analyses is that we do not begin with an overarching theory or model of thought about the elements. Instead, we start with an exciting convergence of thinkers whose works have largely been attuned to elemental phenomena (atmospheres, chemicals, water, soil, fire, wood, and more), even if they had not centered the elements as an analytic in their work. When we invited the contributors to this collection, the brief was open, trustful. The invitation to contribute was made in the spirit of speculative inventiveness, since we knew that all these scholars are committed experimenters. We wanted to see what forms of elemental thought could take shape from attentions already so carefully attuned to lively and deadly materialities, temporalities, and relations of power in our technoscientific worlds. In shaping this project, we appealed to authors who had inspired us, such as Tim Choy's (2011) ethnography of air pollution, Cori Hayden's (2007, 2012) work on the relation between the chemical and the political, Stefan Helmreich's (2009) accounts of oceans and waves, Joseph Masco's (2004) research on radioactive nucleotides and mutant ecologies, Michelle Murphy's (2017) decolonial analyses of industrial chemicals, Patrick Bresnihan's (2015, 2016) inquiries into the contemporary politics of the more-than-human commons, Astrid Schrader's (2010, 2017) theorization of posthumanist attunements with ambivalent algae toxicities, Joseph Dumit's (2004, 2012) experimental methods for documenting the implosions we call brains and pharmaceuticals, and Isabelle Stengers's approach to the making of science (Stengers 1993) and the practice of chemistry (Bensaude-Vincent and Stengers 1996). Of course, the invitation was driven by some of our own research commitments too—Dimitris Papadopoulos's (2011, 2018) proposition for alterontological technoscientific experimental practices, María Puig de la Bellacasa's (2014, 2019) investigations of more-than-human soil relations, and Natasha Myers's attentions to lively chemistries and their affective ecologies (Hustak and Myers 2012; Myers 2015). And so, our engagements with the elements draw on long-standing conversations on matter and materiality across fields.

In this volume, these influences set the elements into motion to alter our thinking and generate unexpected activations. The effect is an exploration that multiplies claims to what matters in thinking about the elements in theory, method, and practice. We are attuned to conceptions and attentions that keep elemental thought open and responsive to the "ecologies of practice" (Stengers 2010) that shape our multiple, shifting, and overlapping obligations to the worlds we research and for which we write. The contributions involve the elements in different ways: as chemical categories, as cosmological forces, as material things, as social forms, as forces and energies, as sacred entities, as experimental devices, as cultural tropes, as everyday stories, and as epistemic objects. Yet they are also all compelled by the elements' ambivalence as material chemical substances, by their embeddedness in ecological relations, by the relevance of their sociocultural and political actuality, and by their potential to galvanize alternative *practice*. Multidimensional, multiscalar, multisited, multimodal: this volume activates ways of palpating the elemental multiverse.

Throughout this volume, there is also a common, underlying sense of urgency for more attention, appreciation, and care for how human and more-than-human beings, their agencies and collective dynamics, are shaping Earth's possible futures. How can thinking with the elements help us engage with the precarious situation of contemporary life and death in technoscientific worlds? How does a world begin to look when we think it from its elementals and their compositions and decompositions? The elements and their relations always exceed our renderings of them and so push us to develop more subtle analytics about their ongoing material and energetic rearrangements. Who, we must ask, are the "agents of elemental rearrangement" terraforming the earth today (Myers 2017)? Who and what are responsible for composing the elements and

facilitating their decomposition? How are we implicated in the proliferation of damaging chemicals? How can we promote nonviolent elemental relations? In sum, what, in this time of late industrialism (Fortun
2012), are our obligations and responsibilities in mediating the distribution and redistribution of the elements?

The call to wake up elemental thought is, in part, an effort to thwart
the dire thinking perpetuated by Anthropocene logics. To disrupt current imaginaries of the inevitability of apocalypse, we reactivate the
elements as both more-than-human and more-than-natural forces. We
recognize that humans are not the only agents terraforming Earth and
simultaneously that how people rearrange the elements has serious
worldly consequences. The anthropocentrism of Anthropocene rhetoric tends to vault "Man" to the position of ultimate agent and arbiter of
our current predicaments. As many critics charge, this logic continues to
perpetuate the delusion that humans are alone, separate from nature,
and that technological fixes are what will mitigate disaster or provide
the ultimate exit strategy from this damaged planet (see, e.g., Davis and
Todd 2017; Haraway 2015; Myers 2018). In this volume, we reckon with
the ways that the elements are not just matters and forces whose natures
are the domain of science and industry and whose power is beyond the
human. We read the elements not as the nature which "humanity" must
struggle against or tame, but as naturecultures. In this vein, if climate
change can be understood as a rearrangement of the elements at a planetary scale in the aftermath of colonialism, productivism, and their ex-
tractions, then at least we will never again think of the atmosphere as a
nonhuman matter.

Reactivating the elements allows us to think not only against Anthropocene logics that fetishize human primacy but also beyond the
partitioning of species thinking. By drawing on insights from the more-
than-human turn, we seek to go beyond entrenched oppositions between
life and nonlife (e.g., Povinelli 2016) and between substance and process
in order to engage with biogeochemical processes of breakdown, compounding, decay, and corrosion, with an attention to cosmic, cosmological, and cosmopolitical relationalities. By tapping into cosmic, earthly,
and human agencies and attending to the accountabilities, responsibilities, and obligations involved in these reactivations, these texts examine how we might learn to do life and death on this planet otherwise. In
short, these chapters offer expanding views of this moment of elemental upheaval.

The chapters in this book activate various aspects of thinking with the elements. Each contribution bonds with and cleaves to others, linking together diverse questions of substance, ecology, actuality, and practice. Some chapters present research around one element in the style of empirical science and technology studies scholarship, others activate conceptual and methodological interventions, some generate speculative fabulations, and some do all of that in one.

ELEMENTS AS RELATIONAL SUBSTANCE

Today, technoscience dominates thinking around the elements as the *substance* that makes the world. Biology tells us that life is made possible by four organic building blocks: carbon, hydrogen, oxygen, and nitrogen. Chemistry and physics see the elements as the primary constituents of matter, with organic and inorganic worlds ultimately breaking down into hydrogen, helium, iron, and more. For most economists, industrialists, and engineers, the elements are resources that are isolatable, combinatory, and divisible; they are raw materials for warfare, electricity, agriculture, pharmaceuticals, and the bioeconomy. The elements are mattered as extractable commodities, alienable bits of life and death and nonlife, and as energies that fuel global economies. They are put to work in large-scale state and industrial infrastructures—from dams to wind farms. And those consigned to extract value from the elements must labor in high-risk jobs inside industrial plants synthesizing chemicals, in oil and gas wells extracting fossil fuels, and in mines harvesting valuable minerals. For those who labor, and those who live downstream from the toxic ecologies of late industrialism, the elements and their rearrangements are simultaneously hazards, harms, and hopes.

Reactivating Elements opens with an introductory intervention by Isabelle Stengers, who has transformed the way we think about what the sciences mean and what they do by calling us to both value and be wary of their specificities and powers. We learn from her contribution why we should resist any purified definition of elements. Stengers also offers a critical history of the elements of chemistry that attempts to recognize their specificity and at the same time refuses to think of them in isolation from other naturecultural reactivations in technoscience. She addresses ideas that resonate throughout this collection, inviting contemplation about compositions and decompositions, relational affinities and invariants. Her chapter beautifully situates the overall project of this collection as an exploration of "agency as freed from the opposition between living intentionality and the inorganic." She approaches elements as "metamorphic": "both shaping and being shaped by the particular ecology in which they participate."

"What is an element capable of?" is a key question that structures Dimitris Papadopoulos's chapter. The ecologies in which chemicals emerge or are manufactured and their presence in them cannot be separated. By engaging with different registers of ecology and envisioning chemistry beyond science as "chemical practice," Papadopoulos sketches a nonreductionist way of engaging with the world from the perspective of chemicals, an approach that tries to avoid the binarism between a critique of chemical horrors and the celebration of their powerful achievements—the two dominant poles of engagement with chemical pervasiveness. His contribution asks fundamental questions for our chemically compelled times: What practical obligations could emerge from a nonbinary approach to the ubiquity of anthropogenic chemicals? And what are the meanings of ecological reparation and justice within this context?

In chapter 1, Stefan Helmreich approaches elements as organizations of matter, energy, and life and puts them in discussion with the irreducible "elementary forms" that animate the history of social theory. He looks at resonances between the ways Mendeleev found a logic to drive his periodic table of elements and the way Durkheim defined "the social" as an "elementary form" of organization with a similar ambition: "one that posits social organization as a composing logic." In turn, twenty-first-century elemental thinking refuses clean and clear divisions between science and society, nature and culture, the inorganic and the organic. For Helmreich, this is "an attention to hybrids, chimeras, and material-symbolic mixtures—as well as their inextricable multispecies politics." Helmreich's reactivation of water and waves inherits both old and new elemental thinking and invites us to rethink water and its waves as an "elementary form" of both social and elemental life.

Joseph Dumit's chapter makes salient what is implicit in the three first chapters: elements as relational substances. Driven by passion for the specificity of a substance, Dumit believes that "each substance is elemental." He invites us to consider bromine, a highly reactive element, through the ways that it surprises (or reactivates) the specialists who study it. Crucially, he attends to method and asks whether there are "processes that keep us enthralled to our existing concepts" that are "keeping us from seeing more"? He looks to bromine's activities in his multiple research contexts, from its role in the fascia that holds our bodies together, to its function in drug design, to its toxic effects in fracking fluid. The chapter becomes an exploration of the unknowns that can unsettle researchers' methods and concepts. By studying substance through its effects and its affects, we learn not only about bromine but also about how substances put the vocabulary of researchers "into variation." When Dumit suggests that a "substance can become a method of undoing one's theoretical assumptions," we are invited into another dimension of reactivation that opens our thinking and theories to change as we encounter different substances.

ACTUALITIES OF THE ELEMENTAL

The adventurous empiricism of Dumit's bromine investigations leads to the following four chapters, which engage elements in their actuality, asking how elements are in the making and how they are always tightly imbricated to earthly ecologies and to wider cosmic and human powers. Papadopoulos reminds us of Whitehead's rendering of a molecule as a "nexus of actual occasions" (1979, 73) and "a historic route of actual occasions." This can also be said of elements. Different ecologies, different cultures, orders, economies, and matters activate different distributions of elementary forces. Therefore, they are always already entangled in local ecologies and in historically and geographically situated social, cultural, political, and economic ways of doing life.

Astrid Schrader's chapter is a plea for a more-than-human approach to agency that takes actuality seriously. Schrader engages with debates in geography, social theory, and science and technology studies in concert with elemental thought to investigate how viruses mediate the biogeochemical rearrangements of the oceanic carbon cycle. Engaging contemporary ecocriticism's reactivation of elemental theory, Schrader reframes the elements as partners in world making, challenging entrenched distinctions between life and nonlife. For her, viruses are forms of life-nonlife that she refigures as "elemental ghosts," not as sensible things but "of the sensible." Taking viruses as a guide, she opens up a mode of thought that pushes past ingrained assumptions about immutable ontologies and into a way of engaging elemental substance as process and relation. Here, the elemental activates an exploration of a nonanthropocentric carbon imaginary that extends beyond the sociopolitical realm and also cuts through the dichotomy that entrenches the inertness of the geological in opposition to the liveliness of the biological.

In the next contribution, Joseph Masco starts with a question that reactivates elements thinking in all its actuality: "What is a world actually made of?" Inquiring about "a" world and not "the" world tempers any absolutist inquiry into the elemental and the quest for elemental origins. Contrasting the orderly portrait of the world offered by the chemists' periodic table with the messiness and complexity of actual ways of living lays the ground for an exploration of the materials that have become sociotechnical signatures of the Anthropocene. Masco calls these substances "ubiquitous elements": the artificial compounds that nuclear war and mass production have distributed across the planet. The chapter explores the technoscientific, political, and affective history of two very different ubiquitous substances: the new eternals, which are the most fetishized elements (plutonium and other radioactive elements) and the indecomposables, which are the most mundane materials (plastic). Masco makes these familiar compounds strange, showing how these actualities were consolidated in mid-twentieth-century industrial and military achievements radiating from the Global North, and how they thus became an ontological condition on planet Earth, turning one time into deep time.

Patrick Bresnihan's chapter opens with a vivid description of the effects of wind on people and land on the west coast of Ireland, inviting readers to feel the presence and experience of wind through culture, myth, and poetry. This immersion reactivates the material semiotic power of the weather's elemental forces. Wind has acquired new historical roles with climate change and the intensified elemental forces it has unleashed (e.g., Boyer 2019). Wind now labors for capitalism, enrolled as a form of "work/energy" to drive green economic growth. Wind as resource is a "new frontier." Bresnihan's analysis makes salient a thread that runs through this collection, perhaps made more explicit by his search to both counter the economization of wind and seek alternative readings and practices. Bresnihan scrutinizes attempts to harness wind in Ireland and reveals how industrial efforts to capture wind have cascading destructive effects. Then, in contrast to dreams of "smart" ecological modernization, he offers a "speculative antidote" to wind farms as a way to "re-enchant wind" by engaging nonscalable wind projects that refigure it beyond "anthropocentric interest" and as a potentially unwilling collaborator in lively, precarious, and more-than-human experiments.

Cori Hayden shifts our attention from blustering winds to swarming crowds with a chapter examining the elementary forms of social life that generate the "collective effervescence" associated with crowds' irreducible dynamics. What are crowds made of? What makes them hold together? But also, what does it mean to think with crowds today? Exploring how crowd theory has changed between the nineteenth century and today, Hayden shows how elemental thought, complete with its chemical and physical metaphors, is deployed to describe social behavior. Crowds are seen to behave like waves, fire, and wind, dissolving the individuated liberal subject into swarms of energy, "imitation rays" and magnetizations that override individual drives, propelling social movements, both liberatory and repressive. Assertively addressing the entanglements of crowd theory with early twentieth-century fascism, racism, and elitism, Hayden examines how crowd theories "vividly recompose social theory's vocabularies for parts, wholes, and the ties that bind." We learn how nineteenth-century crowd thinking—led by social theorists such as Gabriel Tarde and Gustave Le Bon—diverges from liberal democratic thought that locates the elementary forms of social action in the rational individual. And yet, as social media swarms and crowds assert their power to reshape democratic processes, Hayden looks to crowd theory's efforts to "recompose the elements of social action and analysis" as a way to remind us of both the political force of elemental thought and the illiberalism endemic to liberal democracies.

ELEMENTAL PRACTICES

Against the background of agitated and urgent questions around environmental politics and the future of life on Earth, the last three contributions to this volume call upon a form of elemental thinking that keeps our analyses grounded in relations of power. Who and what have the power to rearrange the elements, to terraform Earth and alter its atmosphere? Colonial conquest, the industrial revolution, extractive capitalism, and late industrialism have all staged particular relations with the elements. Thinking with the elements can help us learn how to describe what Michelle Murphy (2017) calls "alterlife's chemical relations"—its toxicities, harms, and extinctions. Yes, the catastrophe is now. And yet we need to get beyond descriptions of our dire present to dare to dream of alternative practices within these worlds. We cannot forget that for many working intensively to resist the toxic legacies of industrialism—including Black and Indigenous communities, and communities of color, ecofeminists, neopagans, and ecological activists—the elements retain sacred, relational, naturecultural potentials. Thinking with the elements may be one way to wake up the alchemical, creative agencies of the cosmos and resist efforts to enclose them as resource and commodity. How do we learn to live with and work against industrialism's elemental aggregations, assemblages, arrangements, and rearrangements? Is it possible to rearrange our relations with the elements to activate what Papadopoulos (2018) calls "alterontologies"? How can reactivating the elements help us attend to the making and unmaking of worlds? The last three chapters take what María Puig de la Bellacasa calls "ecopoethical license," fusing ecology, poiesis, and ethics to practice elemental forms of speculative fabulation. Enlisting creativity and invention in service of activism, these contributions engage artmaking, artworks, and artists to move beyond the text and to catalyze new questions about our obligations and responsibilities to elemental relations. Here the elements reactivate us as practitioners of alterlife, as co-conspirators with the elements in efforts to resist harm and violence, while attempting to transform earthly processes of living and dying.

Thinking with soils, as proxies for the earth element, María Puig de la Bellacasa proposes that embracing the breakdown of matter is a way to create alternative paths that respond to the ways technoscience chokes up the biogeochemical cycling of elements. She engages with the ethico-political dimensions of a contemporary ecocultural reactivation of elements, one promoted by an imaginary of planetary elemental affinities, fed by scientific storytelling and ecological aesthetics, as well as by political theories of the elemental commons. Contesting the powerful cultural imaginary that equates life with growth and attainment, Puig de la Bellacasa engages with soils through the biochemical processes of breakdown, decompounding, and decreation. Meditating on the ethical, poetical, and poietic significance of embracing breakdown for the ongoing reshaping of "the ecological," her "panethnographic" research invokes accounts of "bioremediation" as a noninnocent, practical enactment of "breakdown ecopoethics" to contribute alternative stories of more-than-human relations that might not only celebrate common material earthly affinity but also activate solidarity with the present struggles of soils in their actuality—neglected, sealed, and contaminated.

In Tim Choy's chapter on atmospheres, air's agency is embraced as a more-than-human arrangement with those who breathe it. Choy's chapter explores the "viscerality" of the concept of "breathers" as those beings who must "pay" for the unpaid costs of capitalism, the market externalities that accumulate in the airs we breathe. Choy emphasizes the "unequal shared cosubjection" of breathers in a disrupted atmosphere and uses the creative form of comic strip style drawings as a mode of speculative experiment to provoke a kind of atmospheric reckoning of our intimate relations with air and with one another. He mobilizes the concept of conspiracy, whose root meaning is to *con-spire*—that is, to breathe together—to imagine a *conspiracy of breathers* as a political formation that might resist the industrial drive to pollute. His drawing practice thus becomes part of "a speculative project of collectivizing response to massive, patterned forms of environmental violence" and a method to transform the ways we may account for this loss. His remarkable drawings in the "Museum of Breathers" amplify the more-than-human interdependencies that can compel us to respond, proposing we become accomplices in a co-conspiracy that can rearrange relations in this unequally shared atmospheric milieu.

Reactivating Elements ends with a contribution by Michelle Murphy on the "alterlife" of persistent organic pollutants in and around the Great Lakes, in the aftermath of the "industrial exuberance" of Chemical Valley—an industrial corridor built on stolen Indigenous land that is responsible for more than 40 percent of Canada's petrochemical processing. Murphy tracks the flows of rainfall that wash the PCBs (polychlorinated biphenyls) accumulating on urban skyscrapers out through stormwater sewers into Lake Ontario, swirling in great eddies of contamination and enduring exposure. She shifts technoscientific genealogies from the mainstream critics of science to Black and Indigenous thinkers, artists, and activists leading decolonial movements. Murphy identifies state-sanctioned permissions to pollute as elements of a colonial project and exposes corporate strategies that create "infrastructures of not knowing" to avoid even the minimal self-reporting of industrial emissions that governments demand. She calls this infrastructural gaslighting and invites us to focus on documenting and studying colonial violence rather than damage narratives that tend to put "the burden of representing violence on already dispossessed communities." In the mode of speculative fabulation, she ends the chapter with an invitation to visualize an expansive iconography for industrial chemicals that can account for their extended relations of harm and violence.

These final chapters vividly express another thought perceptible throughout the whole collection: a call for experimental practices committed both to collectively exploring alternative ways of experiencing environmental destruction and to activating potentials for alternative elemental rearrangements. And so, going beyond complacent critique, *Reactivating Elements* engages with elements in diverse actual occasions, methodological configurations, and practical involvements, experimenting with forms of storytelling, concepts, and methods with the hope of triggering possibility and a sense of obligation toward worlds we participate in making. Throughout this book we are watchful for how elemental thinking can offer new ways to make sense of, and to care for, the naturecultures we are actively and continuously worlding.

NOTE

1. This collection started with a panel titled "Elements Thinking" at the 4S/EASST conference "Science and Technology by Other Means: Exploring Collectives, Spaces and Futures," August 31–September 3, 2016, Barcelona, Spain.

REFERENCES

Alaimo, Stacy. 2016. *Exposed: Environmental Politics and Pleasures in Posthuman Times*. Minneapolis: University of Minnesota Press.

Barad, Karen. 2007. *Meeting the Universe Halfway: Quantum Physics and the Entanglement of Matter and Meaning*. Durham, NC: Duke University Press.

Bennett, Jane. 2010. *Vibrant Matter: A Political Ecology of Things*. Durham, NC: Duke University Press.

Bensaude-Vincent, Bernadette, and Isabelle Stengers. 1996. *A History of Chemistry*. Cambridge, MA: Harvard University Press.

Boyer, Dominic. 2019. *Energopolitics*. Durham, NC: Duke University Press.

Braun, Bruce, and Sarah Whatmore, eds. 2010. *Political Matter: Technoscience, Democracy, and Public Life*. Minneapolis: University of Minnesota Press.

Bresnihan, P. 2015. "The More-Than-Human Commons: From Commons to Commoning." In *Space, Power and the Commons*, edited by S. Kirwan, L. Dawney, and J. Brigstocke, 93–112. London: Routledge.

Bresnihan, P. 2016. *Transforming the Fisheries: Neoliberalism, Nature and the Commons*. Lincoln: University of Nebraska Press.

Choy, Timothy K. 2011. *Ecologies of Comparison: An Ethnography of Endangerment in Hong Kong*. Durham, NC: Duke University Press.

Clark, Nigel. 2011. *Inhuman Nature: Sociable Life on a Dynamic Planet*. London: SAGE.

Cohen, Jeffrey Jerome, and Lowell Duckert. 2015. *Elemental Ecocriticism: Thinking with Earth, Air, Water, and Fire*. Minneapolis: University of Minnesota Press.

Coole, Diana H., and Samantha Frost, eds. 2010. *New Materialisms: Ontology, Agency, and Politics*. Durham, NC: Duke University Press.

Davis, Heather, and Zoe Todd. 2017. "On the Importance of a Date, or, Decolonizing the Anthropocene." *ACME: An International Journal for Critical Geographies* 16 (4): 761–80.

Dumit, Joseph. 2004. *Picturing Personhood: Brain Scans and Biomedical Identity*. Princeton, NJ: Princeton University Press.

Dumit, Joseph. 2012. *Drugs for Life: How Pharmaceutical Companies Define Our Health*. Durham, NC: Duke University Press.

Ernstson, Henrik, and E. Swyngedouw. 2019. *Urban Political Ecology in the Anthropo-obscene: Interruptions and Possibilities*. London: Routledge.

Fortun, Kim. 2012. "Ethnography in Late Industrialism." *Cultural Anthropology* 27 (3): 446–64.

Haraway, Donna. 2015. "Anthropocene, Capitalocene, Plantationocene, Chthulucene: Making Kin." *Environmental Humanities* 6: 159–65.

Haraway, Donna. 1988. "Situated Knowledges: The Science Question in Feminism and the Privilege of Partial Perspective." *Feminist Studies* 14 (3): 575–99.

Haraway, Donna. 1997. *Modest_Witness@Second_Millennium: FemaleMan©_Meets_OncoMouse™; Feminism and Technoscience*. New York: Routledge.

Hayden, Cori. 2003. *When Nature Goes Public: The Making and Unmaking of Bioprospecting in Mexico*. Princeton, NJ: Princeton University Press.

Hayden, Cori. 2007. "A Generic Solution? Pharmaceuticals and the Politics of the Similar in Mexico." *Current Anthropology* 48 (4): 475–95.

Hayden, Cori. 2012. "Population." In *Inventive Methods: The Happening of the Social*, edited by Celia Lury and Nina Wakeford, 173–84. London: Routledge.

Helmreich, Stefan. 2009. *Alien Ocean: Anthropological Voyages in Microbial Seas*. Berkeley: University of California Press.

Hustak, Carla, and Natasha Myers. 2012. "Involutionary Momentum: Affective Ecologies and the Sciences of Plant/Insect Encounters." *Differences* 23 (3): 74–118.

Kimmerer, Robin Wall. 2013. *Braiding Sweetgrass: Indigenous Wisdom, Scientific Knowledge and the Teachings of Plants*. Minneapolis: Milkweed Editions.

Lefèvre, Wolfgang, and Ursula Klein. 2007. *Materials in Eighteenth-Century Science: A Historical Ontology*. Cambridge, MA: MIT Press.

Macauley, David. 2010. *Elemental Philosophy: Earth, Air, Fire, and Water as Environmental Ideas*. Albany: SUNY Press.

Masco, Joseph. 2004. "Mutant Ecologies: Radioactive Life in Post–Cold War New Mexico." *Cultural Anthropology* 19 (4): 517–50.

McGregor, Deborah. 2014. "Traditional Knowledge and Water Governance: The Ethic of Responsibility." *AlterNative: An International Journal of Indigenous Peoples* 10 (5): 493–507.

Murphy, Michelle. 2017. "Alterlife and Decolonial Chemical Relations." *Cultural Anthropology* 32 (4): 494–503. doi: 10.14506/ca32.4.02.

Myers, Natasha. 2015. *Rendering Life Molecular: Models, Modelers, and Excitable Matter*. Durham, NC: Duke University Press.

Myers, Natasha. 2017. "Photosynthetic Mattering: Rooting into the Planthroposcene." In *Moving Plants*, edited by Line Marie Thorsen, 123–29. Næstved, Denmark: Rønnebæksholm.

Myers, Natasha. 2018. "How to Grow Livable Worlds: Ten Not-so-Easy Steps." In *The World to Come: Art in the Age of the Anthropocene*, 53–63. Gainesville, FL: Samuel P. Harn Museum of Art Press.

Neale, Timothy, Thao Phan, and Courtney Addison. 2019. "An Anthropogenic Table of Elements: An Introduction." *Fieldsights*, June 27, 2019. https://culanth.org/fieldsights/an-anthropogenic-table-of-elements-an-introduction.

Papadopoulos, Dimitris. 2011. "Alter-ontologies: Towards a Constituent Politics in Technoscience." *Social Studies of Science* 41 (2): 177–201.

Papadopoulos, Dimitris. 2018. *Experimental Practice: Technoscience, Alterontologies, and More-Than-Social Movements*. Durham, NC: Duke University Press.

Parikka, Jussi. 2015. *A Geology of Media*. Minneapolis: University of Minnesota Press.

Peters, John Durham. 2015. *The Marvelous Clouds: Toward a Philosophy of Elemental Media*. Chicago: University of Chicago Press.

Povinelli, Elizabeth. 2016. *Geontologies: A Requiem to Late Liberalism*. Durham, NC: Duke University Press.

Puig de la Bellacasa, María. 2014. "Encountering Bioinfrastructure: Ecological Struggles and the Sciences of Soil." *Social Epistemology* 28, no. 1: 26–40. doi: 10.1080/02691728.2013.862879.

Puig de la Bellacasa, María. 2019. "Re-animating Soils: Transforming Human–Soil Affections through Science, Culture and Community." *Sociological Review* 67 (2): 391–407. doi: 10.1177/0038026119830601.

Robbins, Paul. 2012. *Political Ecology: A Critical Introduction*. 2nd ed. Oxford: Wiley-Blackwell.

Rose, Deborah Bird. 2004. *Reports from a Wild Country: Ethics for Decolonisation*. Sydney: UNSW Press.

Schrader, Astrid. 2010. "Responding to Pfiesteria piscicida (the Fish Killer): Phantomatic Ontologies, Indeterminacy, and Responsibility in Toxic Microbiology." *Social Studies of Science* 40 (2): 275–306.

Schrader, Astrid. 2017. "Microbial Suicide." *Body and Society* 23 (3): 48–74.

Simpson, Leanne Betasamosake. 2017. *As We Have Always Done: Indigenous Freedom through Radical Resistance*. Minneapolis: University of Minnesota Press.

Starosielski, Nicole. 2015. *The Undersea Network*. Durham, NC: Duke University Press.

Stengers, Isabelle. 1993. *L'invention des sciences modernes*. Paris: La Découverte.

Stengers, Isabelle. 2010. *Cosmopolitics I*. Minneapolis: University of Minnesota Press.

Tsing, Anna. 2015. *The Mushroom at the End of the World: On the Possibility of Life in Capitalist Ruins*. Princeton, NJ: Princeton University Press.

Whitehead, Alfred North. 1979. *Process and Reality: An Essay in Cosmology*. New York: Free Press.

1

RECEIVING THE GIFT: EARTHLY EVENTS, CHEMICAL INVARIANTS, AND ELEMENTAL POWERS

Isabelle Stengers

CHEMISTRY'S SPECIFICITY

While preparing my essay for this collection, a personal memory came to be reactivated. As a young first-year student in chemistry, I was asked at an exam, What is an element? It was a turmoil in my head. I knew somehow that to answer "an atom"—as suggested by the familiar models of molecules, colored balls linked by rigid rods—would be to fall in the trap. I knew elements were classified in the Mendeleev table, but what is classified? What does an element mean for the chemist? I never forgot the answer when it was finally given to me: "An element is a chemical invariant."

Some may recall Alan Sokal's *Lingua Franca* paper where he revealed his previous paper published in *Social Text* to have been a hoax or a spoof, and most notably his first derisive comment. I have long wondered why this very paper was not rejected by *Lingua Franca* as a spoof right from the beginning, when Sokal writes: "In the second paragraph I declare, without the slightest evidence or argument, that 'physical "reality" [note the scare quotes] . . . is at bottom a social and linguistic construct.' Not our *theories* of physical reality, mind you, but the reality itself. Fair enough: anyone who believes that the laws of physics are mere social conventions is invited to try transgressing those conventions from the windows of my apartment. (I live on the twenty-first floor.)" (Sokal 1996, 62).

The concluding and seemingly conclusive final challenge—which has been repeated by other science warriors—might indeed figure as a parody of the physicist unable to take seriously the scare quotes he had previously used to play the part of "the physicist converted to cultural studies." For him, physical reality, reality as understood by physicists, is "reality itself," in the sense affirmed by our cry "Don't jump out of this window!" It would seem that before Galileo and Newton, humans—or dogs—were desperately confusing windows and doors. And typically, he forgets that no physicist ever "proved" that what is heavy falls if free to do so—at least in a frictionless medium (if not, a bird will gladly launch itself from the window of Sokal's apartment, and a piece of paper plane out of it by windy weather). The great event Sokal invokes indeed belongs to the reality of the lab, as a place where one can address approximately frictionless movement and obtain "facts" able to answer questions related to the way this movement varies, depending on the conditions imposed by the physicist. The (common) fact of the fall was never the answer to a question. Galileo's question was addressed to the way the velocity of a falling body increases. Again, common knowledge agreed that the body gains speed. But Galileo's facts were able to answer a question which nobody before him had asked: Is the gain to be related with time or with space? It is a seemingly strange question, since the time during which a free-falling body has fallen and the altitude it has lost during this fall seem two inseparable aspects of the same movement. I will not give the answer, because it does not interest the skeptic who falls from the window, but physical reality is the reality for which this answer was decisive, opening the future of the laws of physics.

Typically, the physicist Sokal insults his own science's specificity, probably unwillingly so, lazily resting on the mattress of popularizing mediations which aim at convincing "the public" that there is "one world to discover" and that it is the job of physics to do so. But I learned the importance of having been educated as a chemist before becoming a philosopher through the contrast between the physicist elision of "physical" when reality is concerned and the insistence of "chemical" when the definition of the element is concerned. The mistake I would have made if I had answered that the element is another name for what is indeed an atom would have been to use a physicist's sword to cut through the historical knot of distinctions laboriously constructed between chemists and what they operate on and with.

Those distinctions may be derided by Alan Sokal as social and linguistic constructs indeed, but *social* here actively includes dealing with what chemists would never doubt has a reality of its own. It burns, explodes, poisons. It can be controlled but only with care, and the cry "Do not mix these at room temperature!" may well express a historically acquired knowledge, not a knowledge older than humans, but one which claims a "realist" value shared by all chemists. The way they name and characterize the "these" that would explode if mixed at room temperature may be a "convention" upon which chemists have come to agree, but the danger is not.

Correlatively, the history of chemistry may well be marked by claims of rupture from the past, mimicking physics, but it remains dominated by a story of gestures and procedures that have to be learned, rather than by theories. Since the eighteenth century's initial attempts to understand chemistry in accordance with physical (i.e., Cartesian or Newtonian) reality to the contemporary quantum chemical atoms and molecules, the same contrast has been registered again and again, often with frustration by physicists and with some mocking irony by chemists facing this frustration: "In physics, you understand; in chemistry, you have to learn." Also, each time novel techniques are designed, they offer not only new possibilities of questioning and manipulating but also new occasions for surprises, unfolding a new need to negotiate the relation between what chemists consider that they understand and what they have to accept.

For chemists, the term *technoscience* does not, as it does for physicists like Sokal, convey a critical connotation, emphasizing the self-delusion of scientists who claim that theirs is a disinterested effort to reach an understanding of "reality as it is" and that techniques are only a means toward this goal. For Sokal, just to speak of "physical reality," of the "reality of the physicists," was already an offense, announcing a relativist downgrading of laws that should be recognized not as laws of physics but as laws of reality. But chemists, since the beginning of this science, never stopped to relate their claims to their means. And there is not a touch of sad disillusionment in their case. That they have no access to some "reality in itself" will never be synonymous with the skepticism which envisages a "mute" reality on which scientists would project the categories of their own operations: they know only too well how recalcitrant and full of surprise what they address may be to doubt that it has the power to situate them, rather than the inverse. This is why it is so very important that *Reactivating Elements* carefully avoids extracting a definition of elements "purified" from *technè* and resolutely stands with Haraway's natureculture proposition.

WHAT, THEN, IS A CHEMICAL ELEMENT?

The perplexity I felt when asked "What is an element?" is, I now realize, rather true to a history which is not one of progressive elucidation but a wandering one, the now stabilized answer being obtained by a typical natureculture achievement: elements lend themselves to classification. Indeed, ever since their Greek origin, the definition of what we call elements was marked by contention. Aristotle's definition in *On Generation and Corruption* was not an answer; rather, it was the installation of the puzzle of chemical composition. Aristotle proposed that in a chemical mix, the ingredients cease to exist actually, but they continue to exist potentially, as they can be separated out again. Already, human intervention—that is, separation—interferes in what should be clean-cut categories. It imposes perplexity: What can no longer exist "actually" while still existing "potentially"?

In fact, even the word designating the ingredients entering into chemical composition is perplexing. The Greek name for element was *stoicheion*, and it first designated the engraved lines used to evaluate the position of the shadow cast by a gnomon, that is, the position of the sun in the sky. Stoicheia make it possible to decompose the continuous movement of the shadow into time intervals, "hours"—a typically naturecultural achievement, since the sun is turned into a piece of a time-giving device. In contrast, when Empedocles speaks of "four elements," he uses the word *rhizomata* (roots)—that is, generative principles. Plato and Aristotle articulated stoicheion and *archè* (principle) in their critical comments of pre-Socratic philosophers—but they did not clarify this articulation since they were pursuing other kinds of principle. However, when Aristotle faced directly the question of the chemical mix, the mix made his principles stutter.

In fact, Plato had already devised the master metaphor which articulated stoicheia with generative principles. Stoicheia are like letters, the characters which generate a word. But again, the metaphor is undecidable. Do letters generate a word? Or can they be reduced to some kind of building blocks?

When Lavoisier,[1] in the preface to his *Traité élémentaire de chimie* (1789), claimed that elements are what the chemist cannot decompose, he claimed that he was founding a chemistry that was at last modern, purified from metaphysical speculations. Rather, he was giving the maximum importance to his own work: he had refuted the ancient theory of the four elements by showing that he was able to compose and decompose one of them, water. But what he proposed was also a triumphant version of what Robert Boyle had argued in 1661 in his *Sceptical Chemist*: elements, which cannot be "resolved" into other substances and into which those substances can be decomposed, have nothing particular about them. Their difference is purely factual, related to our means of decomposition. They have the power neither to claim being ultimate building blocks nor to explain the generation and differentiation of chemical substances. Boyle did not express a general, epistemological skepticism but a situated one, that of a practitioner confronting the question of the reliability of the tools he has to use, of their capacity to warrant the interpretation given to their operation.

In 1789, Lavoisier was better equipped than Boyle, and skepticism gave way to the epic story of true science destroying the authority of tradition. However, his operational definition of the element was not able to mute the question traditionally associated with elements as entering into composition—how to understand what causes the substances they compose to hold together or to be decomposed. And with this question comes the specificity of chemistry as a technoscience—that is, its symbiotic codependence with technical processes and procedures. Chemists do not "obey nature," as in Francis Bacon's motto, and nature does not obey them in return. Rather they artfully work with the possibilities of operations which nature makes available—for example, this substance is able to decompose that substance but not this other one. If chemists are masters, it is in "getting nature to behave" according to their aim. Chemists do not operate chemical transformations. They harness the power of diverse, selected "agents," but those agents operate on their own, specific terms. This is why many chemists would resist the kind of uniform causation proposed by Newtonian or Cartesian physics. Heat and water may well be mechanical agents, which the chemist needs in order to make a chemical operation possible, but they do not cause it, while chemical agents are causes, but theirs is the always-specific power to decompose and enter into new compositions. Such a power was generally characterized in terms of affinity. A chemical decomposition might be compared to a kind of betrayal, as illustrated by Goethe in his *Elective Affinities*: a (marital) composition will be destroyed and a new one created with an intruder if there is a stronger affinity between one of the partners and the intruder than there was between the two original partners.[2]

But what were the substances I cautiously referred to as partners? The title of the first version of Mendeleev table, published in 1869, was "An Attempt at a System of the Elements Based on their Atomic Weight and Chemical Affinity." An order had at last been discovered, which described elements as principles, divorced from any substantial mode of existence. Oxygen gas is not an element but a simple substance. An element is no longer what cannot be decomposed. What chemists deal with are not elements, but what they obtain can be explained in terms of elements, and the question of affinities, of what chemists operate with but not upon, is to be formulated as properties of the elements. Indeed, the periodic relationship which appears if elements are ordered following their atomic weight concerns affinities—that is, elements that belong to the same class (or column) "behave" in a similar way, that is, have similar affinities for elements of another class.

Mendeleev elements are "abstract," but they are not the product of a cognitive operation of abstraction. They have a real existence, not a theoretical one. They have individuality, which they retain regardless of chemical transformation, and, strangely enough, the appetitive character associated with the term *affinity* was not so very misplaced.

CHEMICAL INVARIANTS

"Elements are chemical invariants": this sober definition, which eluded me when I was a first-year student, was not that of "skeptical chemists" refusing quantum chemical atoms but of chemists who did hold to Mendeleev's "abstract and real" elements, which keep their individuality throughout chemical transformations, whatever the diversity of the roles they play in the many compositions they enter into. As Primo Levi understood, elements are indeed "characters," but not in the ancient typographic sense of letters or *grammata*, but rather in the sense in which a good novelist makes you understand that the same character who turned into an abject criminal might, in other circumstances and with other associations, have turned into a saint.

"Reactivating the elements" will not inherit the surprising discovery of Mendeleev, that of the tabular order that emerges when elements are arranged according to their "atomic weight." Atomic weights now belong to physicists. They are no longer related to chemical operations but interpreted as deriving from what really matters: the number of protons in the atom's nucleus, which itself determines the number of electrons of its shell. As for chemical agential realism (acids "really" attack), it is just a consequence of the sometimes very complicated structure of the incompletely occupied atoms' outer shells. Quantum chemical atoms' associations, it is said, result in giving to the associated atoms' respective electronic shells a "satisfactory," stable structure. For physicists, elements are part of the folklore of the inhabitants of a territory they have conquered: chemistry is a subaltern science, ruled de jure by quantum laws.

De facto, however, quantum chemistry needs these inhabitants to give relevance to its laws. Atoms, now explaining away elements, have nevertheless kept a distinctly chemical flavor. Of course, all the agential verbs that constellate around this explanation—aiming at, obtaining, getting, taking, giving, exchanging, sharing—will be said to be only metaphors. This is not the language of quantum theory. But this language is not able by itself to make sense of chemical agents. Quantum chemistry is still chemistry, something you have to learn, not understand—that is, derive from a general theory. It may certainly be claimed that the hydrogen atom, with one electron only, can be theoretically understood, as well as its stable association H-H. But for the rest, quantum chemists deal with constructions informed both by theory and the exploration of the versatile character of chemical agents. Theory has to follow and interpret what chemists learn about the metamorphic behavior of the inhabitants of the Mendeleev table.

ELEMENTAL ECOLOGIES

The order disclosed by Mendeleev's arrangement is a very particular naturecultural, technoscientific achievement. It cannot be separated from the transformation of the laboratory from the eighteenth to the nineteenth century. The chemist described by Gabriel-François Venel in 1753, in the article "Chymie" of the *Encyclopédie* of Diderot and d'Alembert, was the practitioner of a craft which he finally characterized, after Becher, as a madman passion, an eccentric, asocially obsessed, health-destroying passion that could devour a life. It indeed took a lifetime to learn the "hunch," the educated intuition a chemist needs. Artificial measuring devices were useless, generalities were misleading: chemists could trust only their hard-won expertise when dealing with the unruliness of what their modern counterparts would call semipurified products, unable as such to act as reliable agents. In contrast, at the time of Mendeleev, chemists needed only a few years to become able to participate in the progress of their science. Theirs has become a deeply socialized activity for which the formulation of common definitions is crucial, ensuring that they can share protocols. They work in highly equipped environments and operate in well-controlled, reproducible conditions. Only well-defined agents are allowed in the procedural theater of chemical operations and circulate, like chemists themselves, between the academic lab and the new chemical industries.

What Papadopoulos (this volume) calls an ecology is indeed the best way to characterize this transformation. Venel testified for a chemistry which was rooted in multiple crafts involving a wide variety of recipes, the reliability of which offered no warrant of reproducibility. And he called for a collaboration between "popular" and "scientific" chemistry. At the time of Mendeleev, craftsmen had long since disappeared from the landscape. The procedures chemists used to obtain their agents, meant to be valid whatever the place, circulated in a dense academic and industrial network, articulating protagonists in need of each other, even if their aims might diverge. It is a different ecology with different interdependencies between different protagonists.

If there must be a reactivation of elements, which at last got their scientific identity card from this productive entangled network, it should unfold in other, very different ecologies. These ecologies might reactivate a colloquial use of the term *element*: unleashed elements, such as devouring flames in California, Greece, and Portugal in the summer of 2018, destructive cyclones, devastating inundations, mudflows, and landslides. The ancient four elements may well regain the wild power which was supposedly put under control and now thrive on the anthropogenic transformation of earth ecologies.

But also unleashed is the toxic power of compounds which were created in scientific-industrial labs but are now everywhere on the earth. Those compounds are not "invariant," nor are most of them impossible to decompose. Rather, as Murphy beautifully relates in her contribution to this volume, they are gifted with a long "alterlife," after their life as mobilized by, or involved in, human meaningful aims is over. When released from service—that is, also, when they are out of control and those who created them do not know how, or do not care, to recall them—they give a new natureculture or natureindustry meaning to the difference between *natural* and *artificial*. Elements are now that which resists "natural" processes of decomposition, as achieved not by chemists but mostly by living organisms, especially by tireless bacteria.

Such a difference is again relative to the means involved, but this difference is now formulated in an ecology where cost-benefit consideration is primordial. Biodegradability means cheap elimination. Biotic agents will do the job for free. But when the unleashed molecules cannot be broken down by those agents, whether because they are "aliens" in the dense, entangled communities of agential beings which constitute body or soil ecology or because they come in such quantities that those communities cannot cope with them, they become "elements," technoscientific creatures escaping technoscientific ecology, entering into a regime of unharnessed operations and playing havoc with earth ecologies.

Plastics, industrial nitrogen fertilizers, and pharmaceutical medicines and pesticides are well-defined molecules like any other, selected from innumerable ones synthetized in chemical labs because of the aims they serve and their capacity to participate in intentional projects. To associate them with elemental powers is not to contradict those definitions but to call for a recognition of the "passionate interdependence" (Papadopoulos, this volume) between the practices which address those agents and the ecology which situates them.

ELEMENTS AS MATTERS OF CONCERN

The reactivation of elements is thus an openly impure operation, not the announcement of a clean conceptual turn directed at a new mobilization of researchers. Thinking and learning with the elemental clearly aim at exploring agency as freed from the opposition between living intentionality and the inorganic. But it is not learning "from" the elemental, extracting new authoritative categories which would arm judgment. It proposes a way to think as situated by boundaries between ecologies, the drawing of which is related to circumstances and concerns.

For instance, the concern associated with the wind as a generally unleashed element, which can nonetheless be harnessed for human needs, may be evoked by a French children's song: "Miller, you are sleeping, your mill is spinning too fast. Miller, you are sleeping, your mill is spinning too strongly." But if circumstances change, if wind is to be mobilized by the global grid, the concern becomes very different (Bresnihan, this volume). As an intermittent resource, it will require a new social, technical, natural global ecology. It will come, just as coal and then oil did (Mitchell 2011), "with its world" (Puig de la Bellacasa 2012), not as a "green," innocent, sustainable resource.

To reactivate the elements is to address their metamorphic character, both shaping and being shaped by the particular ecology in which they participate. Bromine the diatomic molecule, not the element, was my nightmare when, as a rather clumsy chemist apprentice, I was watching the red-brown fumes escape from my reactive mixture and attack the fat used to ensure hermetical glassware connections, with only minutes left before the devil would be out of the bottle. But bromine as enrolled by the ecology of the living does not attack or dissolve. As Dumit (this volume) shows, it surprises researchers, enabling biological structures to challenge the categories those researchers trusted.

Elements, as they were defined by chemists, may be enrolled—and harnessed—by chemists. But in so doing, chemists recognize that such achievements come after and even today lag widely behind those of living beings. Chemical elements are protagonists in exquisitely sensitive biotic arrangements inheriting deep-time stories. We, "breathers," all inherit the extraordinary feat which turned oxygen, unleashed as a waste by our chlorophyllian kin, from a strong poison (its general chemical property is its oxidant power) into an indispensable source of energy. The rather desultory contrast with physics ("Physics is something you understand while chemistry must be learned") turns out to be a tame one when we come to the inventiveness associated with life: the more biologists learn, the more they wonder. The first, and unsurpassed, elemental ecology is the involute web of life ecology with its ultrasophisticated capacity to induce, co-opt, and harness elemental metamorphoses. It could be said that life turns the versatility of the elements' character into a matter of vital import.

But the weaving of the web of life is also and inseparably the invention of a multifold precariousness. Because of their metamorphic possibilities, elemental powers are gifts for life, but they can also act as disruptive poisons, the impact of which may affect whole interdependent ecosystems. Ecotoxicology, which was born with Rachel Carson's 1962 *Silent Spring*, is now following the trails of destruction occasioned by molecules new on Earth but continuously produced by our techno-scientific ecologies. It is no longer DDT only but innumerable molecules which now travel the web of life, from living beings to living beings, and are found everywhere, from the apparently most unspoiled regions of Earth to the fetal environment of our children to be born.

Moreover, ecotoxicology is itself dependent on a juridico-techno-industrial ecology which demands, as a matter of primordial concern, that the responsible molecules be indisputably identified if their "guilt" as toxins is to be recognized. And here two ecologies clash, that of the procedures of the lab and that of the intricate interdependencies which make the living organism. In laboratories toxicity is usually identified through dose-dependent procedures: the higher the dose, the higher should be the effect. But as they accumulate in the body, the action of new molecules, foreign to the web of life, appears now to evade this "honest," by-the-book summons. Their ways as elemental agents creeping in the body's labyrinthine ecology are devious, challenging the ceteris paribus clause upon which technoscientific methods of inquiry depend. All other things can never be supposed equal; no action can explain itself independently of its circumstances; weak doses absorbed during long times may have surprising effects as well as a cocktail of seemingly innocuous molecules. Elemental beneficial or disruptive powers cannot be separated from the ecologies they enter. Therefore, their action is to be characterized in terms not of attributable properties but of circumstantial repercussions, or of diffractive causation. Their way of being real is that of the old Greek *pharmakon*, which demanded care and attention, caution and experience. But they will not satisfy the demands of the industries for a clean attribution of responsibility to the powers they unleashed.

CHEMISTRY'S OBLIGATIONS

Together with technoscience itself, chemistry is then at the crossroad. It is de facto "obliged by ecology" (Papadopoulos, this volume). But obligations may be heard as legal or regulatory, explicitly spelled out, derived from what chemists have learned from their mistakes and are bound to "not do it again." Chemistry is used to that. Along the years, its processes and products have included many such obligations which shape its ongoing relation with industrial production. Green products, including bio-degradability as a value, or bioremediation processes learning the use and value of living agents, are already part of the announcement of an ecologically wise technoscience, the promise of a "good Anthropocene," when man will understand "his" responsibility and repair the mess "he" has created.

Is another sense of obligation possible, which would shape chemistry differently—still a technoscience, of course, but belonging to another ecology than the academic-industrial network? Can "obligation stemming from the acceptance of [a] catastrophe" (Puig de la Bellacasa, this volume) resist the business-as-usual project of substituting "existing products with 'green' ones" (Papadopoulos, this volume)? What would be a (techno)science aiming at the "decolonization of matter"? What would it mean to "stay with the trouble" and not trust our capacity to harness powers of remediation? And finally, why then keep the word *science* at all?

For me, the word *obligation* has always been related to the figure of the idiot I took from Gilles Deleuze (2007). The idiot is the one who slows down and hesitates when all others rush, motivated by what is clearly an emergency. Idiots do not deny the urgency but somehow resist mobilization by the emergency. They are unable to formulate what obliges them to slow down, what makes them feel there is something "more urgent," more important. Scientists are rather obviously not Deleuzian idiots when they rush to publish and actively participate in the industrial valorization and public promotion of what they have obtained. But neither are critical humanity academics, when they untiringly hunt whatever they can prey upon as having the slightest flavor of essentialism or idealism. And, as we know, the whole pressure upon academic research is to breed flexible, opportunistic, and productive researchers—true professionals, and certainly not idiots. Scientists are now learning that their partners do not need reliable results; making do is quite sufficient. They know that the practice, which they furiously defended against critics claiming that what they obtain is a matter of social convention only, is in the process of being eradicated, like so many practices before them.

Idiots lack flexibility, and it is the idiotic dimension of scientific practices which will be eradicated, the way their practice demands its practitioners to feel that there is something more urgent than the imperatives of innovation and economic growth. But what may also be eradicated is their possibility to feel accountable not only for the use of their results but also for the way they address what they try to define. Keeping the word *science* is thus the choice of an idiot at a time when the ecology in which modern sciences participated has turned into an extortive one, making the support of research dependent on promises of innovation and other "breakthrough" contributions to industrial competitiveness. At the risk of being dubbed essentialist or idealist, I have dared to propose that "another science" is possible (Stengers 2018) and to dramatize that what is being destroyed are the "obligations" of scientific practices, the way scientific practices (or, more precisely, those scientific practices which would be destroyed) demand that scientists be willingly obliged by what they address, idiotically resisting "good enough" interpretations.

It is important from this point of view that the term *obligation* implies a dimension of gratefulness. When scientists obtain results which at last satisfy their obligations, their joy is not that of the triumphant conqueror. I would propose that it is rather the joy of the one who has received a gift. Of course, scientists usually celebrate such gifts in terms of what they make possible: publications, financing, credit, and so on. But the very value of what they celebrate is that what they have obtained has not been extorted, is not liable to be reduced to an artifact. The answer testifies that their question has been endorsed as relevant by a noncompliant witness. In contrast with flexible professionals, who are satisfied with good-enough answers, practitioners idiotically obliged by the relevance of their questions might come to claim the possibility of this endorsement not as something they would have deserved, but as an event. As William James would have said, they have jumped into and toward a world, trusting that the other parts would meet their jump (see James [1911] 1996, 230) and they have been given the kind of answer which they hoped to obtain. Honoring this event might enable scientists to connect with different sociotechnical spiritual ecologies, learning to work with other protagonists who cultivate other ways of trusting the world.

This proposition does not imply a figuration of the giver, no more than do such gifts, as when seeds grow their first leaves, when a newborn first cries out, or when infants risk their first vacillating steps—or when, in many civilizations, the prey accepts giving its life to the hunter. Thanksgivings do not need to be addressed to a god, nor to Mother Earth; they belong to the event—they are, I would propose, what witches speak about as "Mysteries" (Starhawk [1979] 1999). I would call such events "earthly events," because they signify not a transcendent power but rather our belonging to and with something greater than us. And, contrary to a (catholic) miracle, a mystery has no need to resist explanations. Explanations, whether religious or scientific, just come after the event, to be spiritually cultivated or relegated to human subjectivity.

Obviously scientific "idiots" do not usually cultivate such events; rather, they inventively devise what might be their consequences. Right from the beginning, when Galileo transformed what he had obtained into a consequent polemical argument, scientists have turned what they achieve into a generally required standard, to be opposed to "irrational" opinions. They have presented the gift they happen to receive, facts that verify the way they address the situation they have crafted, as something they are entitled to demand, whatever the situation they address.

Ungratefulness is dangerous, as it generates dreary arrogance and carelessness, predators who decipher their world in terms of potential preys only. But when it comes to be felt as such, it is shame, rather than guilt, that is experienced. And this matters, as the guilty ones take up all the room, perceive their victims everywhere, feel responsible for everything, while those who feel shame are ashamed before what they mistreated, before the gifts they took for granted. They are humbled by, not responsible for.

It may be that "to accept the obligations stemming from the catastrophe" means to accept, as María Puig de la Bellacasa proposes, that the catastrophe is now, that unleashed elemental powers are now humbling our scientific means. Confronting elemental powers which they helped to unleash, chemists might feel shame, but it is important that they still name themselves scientists and not guiltily give up the gifts they have received.

Remediation, if at all possible, might need shame, but certainly not guilt, in order to participate in reworlding processes. It needs sciences which do not claim for themselves the power to cure or clean up, which do not seek redemption through engaging desperate fights to save emblematic victims but whose practitioners are able to mourn and cry for the dismembering of the world (Starhawk 1990, 30–31).[3]

To attend and *to assist* are recurring words in *Reactivating Elements*. Those are the kinds of words we use for midwives, and they may be relevant for scientists participating in the cultivation of reworlding processes. Midwives recognize that the birthing one is the one "laboring" and that she does it her way, even if she can be helped to cope. Scientists who would learn to attend and assist reworlding processes, and who would feel existentially obliged to the ecology they themselves are embedded in (Papadopoulos, this volume), would be grateful if they are able to facilitate those processes. Knowing that they cannot dream to harness and put them to work, they would inherit something of the idiot—an idiot who insists on what is more important than possible remediation success stories: to be part of a cultural, social, political ecology which thwarts any temptation to reduce the powers of what they address to definitions making them agents serving human purposes.

NOTES

1. Jumping directly to Lavoisier should not be understood as confirming that "serious" chemistry begins with him. The book I wrote with Bernadette Bensaude-Vincent (Bensaude-Vincent and Stengers 1996) may interest those readers who would wish for a thicker, less orthodox perspective.

2. Eighteenth-century chemists usually did not take into account the complementary coupling between the two abandoned partners because their practice, like that of craftsmen, centered around those transformations which permit the obtaining of a particular product—that is, "complete reactions." Nineteenth-century chemistry came to understand the results of chemical reactions in terms of merry-go-round, ongoing exchanges between partners, couples continuously made and unmade. Complete reactions corresponded, then, to the particular cases when one of the couples escaped, by precipitation or as a gas, the reactive medium.

3. For the violent means to which conservationists resort to save a species at the edge of extinction, look at the story of the whooping cranes (Van Dooren 2014).

REFERENCES

Bensaude-Vincent, Bernadette, and Isabelle Stengers. 1996. *A History of Chemistry*. Translated by Deborah van Dam. Cambridge, MA: Harvard University Press.
Boyle Robert. (1661) 2003. *The Sceptical Chymist*. New York: Dover.
Deleuze, Gilles. 2007. "What Is the Creative Act?" In *Two Regimes of Madness*, edited by David Lapoujade and translated by Ames Hodges and Mike Taormina. New York: Semiotext(e): 317–29.
James, William. (1911) 1996. *Some Problems of Philosophy*. Lincoln: University of Nebraska Press.
Lavoisier, Antoine-Laurent de. 1789. *Traité élémentaire de chimie*. Paris: Cuchet.
Mitchell, Timothy. 2011. *Carbon Democracy: Political Power in the Age of Oil*. London: Verso.

Puig de la Bellacasa, María. 2012. "'Nothing Comes without Its World': Thinking with Care." *The Sociological Review* 60 (2): 197–216.

Sokal, Alan. 1996. "A Physicist Experiments with Cultural Studies." *Lingua Franca* 6:62–64.

Starhawk. 1990. *Truth or Dare: Encounters with Power, Authority and Mystery.* New York: HarperCollins.

Starhawk. (1979) 1999. *The Spiral Dance.* San Francisco: HarperOne.

Stengers, Isabelle. 2018. *Another Science Is Possible: A Manifesto for Slow Science.* Translated by Stephen Muecke. Cambridge: Polity.

Van Dooren, Thom. 2014. *Flight Ways: Life and Loss at the Edge of Extinction.* New York: Columbia University Press.

2

CHEMICALS, ECOLOGY, AND REPARATIVE JUSTICE

Dimitris Papadopoulos

THE TOXIC REGIME: ANTHROPOGENIC CHEMICALS, ABANDONING EARTH, AND ECUMENICAL PEACE

If our worlds are unimaginable—perhaps even, in a paradoxical way, unsustainable without chemicals that humans make and use—then what does it mean to live and navigate the toxic regime? The toxic regime is a historical moment where anthropogenic chemicals are so entangled with ecology and society and, therefore, chemical contamination is so pervasive that envisioning cleanup is no longer an option. What are our options then?

Abandoning this planet (or large parts of it) has come to shape, often in unspoken ways, many of the future imaginaries, both popular and technocratic, in Global North societies. In the film *Interstellar* (2014), the planet (seen from/as the United States) has become a dust bowl, where crops are devastated by sandstorms. The planetary environment is so unstable that humans give up on taking care of Earth and abandon it. In this technomasculinist fantasy of outer space colonization, humans leave Earth to set up a series of endlessly fertile space colonies. *Interstellar*-like fantasies capture the climate solution for the 1 percent (not even today's 1 percent, but the future 1 percent) which, by deciding to abandon Earth, simultaneously abandons—as Kim Stanley Robinson's *Mars* trilogy (1992, 1993, 1996) depicted compellingly—any possibility for restoring peace and justice on Earth (for analyses of this work, see Jameson 2000; Leane 2002).

A lasting social peace is probably difficult to conceive for many reasons. But within the toxic regime, it cannot even be envisioned. The toxic regime is a primary source of conflict and war. Elemental destruction, contamination, pollution, climate change, extractivism, and resource depletion render parts of the planet uninhabitable and, in turn, perpetuate social injustice. Since anthropogenic chemicals are deeply embedded in matter and operate on temporal registers that are beyond the human, it seems impossible for societies to revert to the ontological configuration of a nontoxic and conflict-ridden Earth. Anthropogenic chemicals inhabit the world, rather than human life; they inhabit Earth, rather than specific social spaces. They pull the many worlds of the planet violently together through their forceful and often deleterious effects. Their reach is therefore ecumenical, and so should be our approach to them. But *ecumenical* here does not just mean to entertain a worldwide scope. With Rob Wilson (2015), I understand this ecumenical perspective as the *housekeeping on Earth*—ecumenical, *oikouménē*, *oikos*, house, inhabiting the world—instead of the housekeeping of various economic and social orders (see Papadopoulos 2018). Rather than a global social perspective, *ecumenical* is about becoming accountable for our worldly existence: our immediate ecologies as oikos; inhabiting our ecologies. When it comes to anthropogenic chemicals, the approach cannot but be ecumenical. And when it comes to peace within the toxic regime, this must be ecumenical peace and ecological justice, not just social peace and justice. There can be only ecumenical peace; or there will be none.

THE CHEMICALIZATION OF SOCIETY: THE DOUBLE SPIRAL OF ECOLOGY AND CHEMICAL PRACTICE

Ecumenical peace means the worlding of peace and justice as a practice of radical "oikology." And therefore, the only option humans have is to practice chemistry as ecology. I do not mean just the encounter of two scientific disciplines, although this might be important in certain cases.

What I mean is the becoming obliged of chemical practice by ecology. Obligation resembles a double spiral, where ecology and chemical practice change as they gradually grow together and one becomes the condition for the other and enables humans—some humans that is—to make alternative ecological chemospheres.

This chapter envisions possible trajectories of this double spiral: how ecology comes to infuse chemical practice and how chemical practice transforms ecological thought and ecological action. The outcome of this is neither a new scientific disciplinary formation nor a new research agenda. It is the creation of alternative ontologies of existence, alterontologies—the crafting of alternative ways of living that materialize ecological reparation, social justice, and community regeneration through the invention of new forms of sociality and technoscientific practice (Papadopoulos 2018).

When I talk of chemistry in this chapter, I refer to something broader than the science of chemistry; I refer to the practice of chemistry: multiple ways of engaging with material substances that involve the making, production, and use of anthropogenic chemicals on different scales and in different conditions, geographies, and settings. The embeddedness of human-made chemicals in modern societies and in industry is so deep that we can talk of the anthropogenic "chemicalization" of society and the nonhuman world (see Barry 2005; Bensaude-Vincent and Stengers 1996; Fortun 2014). Of course, as a science, chemistry is involved in 98 percent of all manufacturing processes and goods. But anthropogenic chemicals are also entrenched in social and material words through a multiplicity of everyday practices that implicate humans and nonhumans in many different and complex ways that go far beyond research in chemistry, chemical engineering, and industrial production. Chemical practice understood as a multitude of transformative engagements with anthropogenic chemicals is far more widespread than chemistry as science itself.

The epistemic frame of anthropogenic chemicals and the engineering frame of industrial production constitute complementary levels to the ontic frame—that is, substances that make up organic and inorganic bodies. Different chemicals and substances are made to matter in each of these different frames. For example, anthropogenic chemicals can be elements and molecules that are distinct epistemic disciplinary objects (of chemistry, physics, or biology); they can be value-creating matter in manufacturing; they can exist as practical-experiential matter in the everyday lives of humans; or they can be part of larger underlying formations of matter (the elemental). It is a political question how to deal with these different frames and whether to see them in their interconnectedness or to split and juxtapose them. In this chapter I argue that a juxtaposition is not only untenable but also politically problematic. My aim, then, is not to put these frames to work against each other through some external demarcation but to see how the traffic between them complicates their relations and eventually allows them to develop in a direction that helps us engage with the pervasiveness of the toxic regime.

THE MULTIPLE REGISTERS OF ECOLOGY: EPISTEME, EXPERIENCE, ECOSOCIAL MOVEMENTS, AND ECOLOGICAL THOUGHT

The other helix of the double spiral, ecology, is—as much as chemistry—something that goes far beyond its own disciplinary coordinates. Ecology involves the epistemic body of knowledge that has been crucial for establishing it as a science,[1] but in the sense that I use the word in this chapter, it entails many other dimensions that are far more important in nonscientific contexts. For example, ecology involves the everyday experience of rootedness and belonging in our surroundings: the embodied understanding of worldly connections between different beings and environments. This is directly linked to another dimension, ecology as "ecosophy," in the words of Guattari (1995, 91)—that is, as a "generalized ecology" with the articulation of "scientific, political, environmental and mental ecologies" (134; see also Hoerl 2013). Equally, the long tradition of environmental activism and ecosocial movements is another constitutive part of what ecology means today (see, e.g., Bullard 1994; Dillon 2014; McGurty 2000; Murphy 2006). Finally, we can think of the multiple ecological cultures, and eventually ecology, as a method of thinking and inquiry.

Unlike attempts to make ecology mean one thing (usually identifying it with the science of ecology), I see all of these meanings as registers of ecology that enact today—in many different, often antagonistic, ways—various aspects of ecological thinking and practice. In this chapter, I engage these different registers to explore how the double spiral between ecology and chemical practice can unfold and create new possibilities. All registers are important for this ecumenical endeavor: the ontic register, referring to the relations and becomings among beings doing life together; the experiential register (resonances, alliances, and belongings with other living beings); the epistemic forms of ecology as a scientific discipline; the political registers of ecology as a set of social movements; the cultural register; and the economic register, which conceives of ecosystems as distinct worldly productive entities (see, e.g., Krebs 2016). It is through all of these different ecological registers that chemical practices can be configured in specific ways, and, equally, different practices of chemistry can enact specific ecological registers and preclude others.

This chapter is organized along several short sections that approach this relation between ecology and chemical practice from different angles. There is no singular, overarching timeline that connects these sections. Most of them depict incidences of ecologically obliged chemical practice presented as short stories of the varied materials I engage with: policy or scientific papers, science fiction, theory, notes from my fieldwork in green and sustainable chemistry labs, excerpts from my interviews with chemists and chemical engineers, historical accounts of chemistry and chemical manufacturing, images, and films. The code for reading this text is to read the sections as stand-alone pieces while knowing that their distinctive specificity is dependent on other stories in the chapter.

ECOLOGICAL TRANSVERSALITY: ETHOPOEISIS AND INTERSPECIES COOPERATION

The females of the *Bactrocera oleae* lay eggs in the mesocarp of olive fruits. It is by far the most damaging pest affecting olives in Mediterranean countries, which account for more than 90 percent of olive production worldwide. The olive fruit-fly larva feeds in the flesh of the olive and pupates in the fruit or in the soil. Traditional means of control include the application of insecticides on the ground, within olive groves, or indiscriminately by air across larger areas. Both methods have large-scale negative impacts on the environment, other beneficial insects, human consumers, and the product itself.

After the female volatile pheromone was identified and synthesized in the early 1980s, several other methods for controlling the olive fruit fly emerged (see Baker et al. 1980; Haniotakis et al. 1986; Mazomenos and Haniotakis 1981, 1985). The pheromone which functions as a male attractant can be used to lure fruit flies into traps that either contain insecticides or physically prevent the flies from exiting before they drown in the bait solution. The adoption of such methods is labor-intensive and necessitates a change of perspective regarding how olive farming is done and perhaps even what olive farming is.

Dionysis Papadopoulos, a small-scale olive and pistachio grower in the Markopoulo area of Attica in Greece, was one of the first farmers to experiment with alternative ways of olive production. I experienced this as his son. Papadopoulos teamed up with the Greek National Science Foundation and offered his farm as one of the first sites in Greece for testing alternative ways of controlling the olive fruit fly using pheromones (see Mazomenos, Pantazi-Mazomenou, and Stefanou 2002). The effect was not just engagement in a new "ethopoeisis"—the simultaneous making of ethos and ontology, in the words of María Puig de la Bellacasa (2010)—of olive farming. It was the establishment of new ecologies—on almost all of the registers mentioned earlier—through the intensive use of these pheromone-based chemicals. Rather than being a means of controlling other organisms, these chemicals reestablished the relation of several involved species and eventually became a means of interspecies cooperation, ecological transversality, and world building on the Papadopoulos farm.

"THE PLAN IS DEATH": SEMIO-MOLECULES, CHEMICAL ECOLOGY, AND BIOPESTICIDES

In James Tiptree Jr.'s (1975) science fiction short story "Love Is the Plan the Plan Is Death" (1975), a male alien arachnid attempts to defy "the plan," the innate behaviors that govern his reproduction and his often-violent relations to other individuals of his species. But the plan always proves more powerful than his own attempts to defy it. He does the same as other members of his species; he shows the same social behaviors; he feeds on his conspecifics. Even the instant in which he believes that he can free himself from these instincts and experience love becomes a moment that fulfills the plan of reproduction. As he releases his love from the silken bonds that he has created to nurture her, they engage in an act of procreation, and then she devours him.

The identification of the female sex attractant of the silk moth in 1959 by the biochemist and Nobel laureate Adolf Butenandt came after long attempts to identify signaling pathways between animals (see, e.g., Butenandt et al. 1959). There was speculation that a special type of electromagnetic waves or some form of ultrasmell, not detectable by the olfactory systems of human or other animals, was responsible for the mediation between animals and between animals and their environment (see, e.g., Fabre 1912). Rather than attempting to identify "the plan," the aim was to reveal how the plan is communicated. There are many signaling paths between organisms, and one is through chemical means (such as pheromones). The term *chemical ecology* came to designate this subfield of chemistry, which studies interactions of organisms using the chemical substances they produce in their bodies.

What if humans could mobilize this chemical mediation to eliminate supposedly "harmful" organisms? When interactive pathways are used as traps, and when they are mixed with toxins or used in ways which disorient the responding organisms, they become devices for insect control. Pheromones allow the communication of "the plan" to be tweaked, to attract and kill, to offer love and administer death. Pheromones came to be seen as a potential replacement for synthetic pesticides—in other words, biopesticides.

With the invention of pesticides, the environment became an invasion field for the extinction of pests (see Landecker 2016). But in the 1950s, as an invasion field, it demonstrated something no one had previously thought of: that it was fundamentally indeterminable and genuinely uncontrollable. Synthetic chemical pesticides, including DDT, were becoming less and less effective, as pests developed resistance to them, and more and more destructive, as ecologies responded to their presence in unpredictable ways. Biopesticides promised to solve the problem. But instead, biopesticides and synthetic, chlorine-based pesticides supported each other in an "arms race [between humans and pests], in which a cycle of action and reaction between opponents produces an escalating spiral" (Wylie 2012, 63).[2]

ECOCHEMICAL INFRASTRUCTURES: WORLD BUILDING AND AFFECTIVE DOING

Starting from the experimentation with pheromone-based chemicals, Papadopoulos created a local version of agroecological practice. He saw in pheromones the possibility to rearrange the ecological presence of his farm. The application of pheromones in the field meant that every other farm activity needed to change too: designing alternative modes of irrigation; trapping fruit flies and other pests before they reached the olive trees in the surrounding fig groves, vineyards; and pasture fields; creating complex ecological relations and plant cooperation that supported insects beneficial for the olive trees; developing a network of local farmers and a system for exchanging information and practices; exploring new knowledge; and engaging in organic farming and distribution.

Although this transformation began with the use of the farm as a scientific experimental site, the reason for Papadopoulos's involvement was not a particular belief in biochemical science. I know from my many conversations with him and by observing him absorbed in his work that it was not just about knowledge but about an affective doing: it was a direct way to learn other, more intensive ways of being with the trees. Papadopoulos was a tinkerer of worlds, a minor inventor, weaving things together that did not really seem to belong to each other, and pheromones allowed for a different, deeper expression of his adoration of these trees. This minor transformation allowed an alternative ecological infrastructure to emerge from these chemical experimental practices. Pheromones were no longer just semiochemicals but building actors of such infrastructures.

Rather than being communicative devices between insects, pheromones become in this story ontological reorganizers of larger ecological infrastructures that involve the insects themselves, other organisms, plants, committed humans, and the social and material worlds they inhabit. Pheromones in this specific case enabled a nontoxic ecochemical infrastructure to emerge. And the more this ecochemical infrastructure of cooperation matured, the less active the dangerous chemicals became in its making. In a world surrounded by synthetic pesticides and small farmer exploitation, this minuscule infrastructure became less toxic, more just, and more mutual.

ECOLOGICAL CONTINGENCY: MOLECULES AS ACTUAL OCCASIONS AND THE *LONGUE DURÉE* OF ELEMENTS

Within ecological infrastructures of animal, human, and plant interaction, the status of molecules has inessential properties. For example, are pheromones a medium when they act as signals and connectors between individuals of a species in a certain ecology? Can molecules just transfer cues, or do they constitute the very content of the cues and the ways the actors involved relate to each other? Is there a molecular specificity of connectivity? And what about disconnectivity? What is a semiochemical when there is disconnection? What are molecules when, instead of just performing the mediating role they are supposed to have within the logic of "the plan," they engage multiple and very diverse beings and things in new ecological configurations?

The point is that chemicals, let alone semiochemicals, do not exist outside their ecologies. Chemicals exist only in context, and they change as the context changes. They become something else; they have different effects; they disappear and reappear. Chemicals are ecologically contingent by definition. If, as Whitehead says, a molecule is a "nexus of actual occasions" ([1929] 1979, 73) and "a historic route of actual occasions" (80), then molecules and the ecologies in which they exist are folded into each other through the route of events that allowed a molecule to exist. Their ecological presence is, of course, multiple: for example, semiochemicals exist as disciplinary objects of science that render them as signaling molecules and, simultaneously, they exist as substances within the ecological spaces in which organisms live (see Hustak and Myers 2013). These two different registers of ecology do not coincide, but they are linked in many different ways; sometimes they are tightly interconnected, and sometimes they are far apart. But ecological contingency is not an epistemological question, so this distinction is irrelevant here; rather, it is about practical ways of materially constructing the worlds that humans and nonhumans inhabit.

Think of oxygen. Two and a half billion years ago, in the Precambrian Period—the "age of chemistry," as it is called—cyanobacteria flooded Earth with oxygen. Together with other significant geological and ecological transformations,—this made the evolution of aerobic organisms possible. Oxygen (and life of Earth) cannot be thought of outside this event; its permanence is not the outcome of its inherent qualities but of its embeddedness in specific ecological conjunctures. Or nitrogen: natural nitrogen fertilizers and fixed nitrogen feedstocks were in high demand after the explosion of industrial production in the nineteenth century, but they were also very difficult to create in large quantities. This was the case only until 1913, when the company BASF—within the context of an arms race leading up to the devastating WWI and rampant large-scale industrialism—deployed for the first time the Haber and Bosch method and realized artificial nitrogen fixation that propelled the manufacturing of synthetic nitrogen fertilizers on an industrial scale. Fixed reactive forms of nitrogen flooded the planet, with innumerable cascading consequences. Even more than history, elements have a *longue durée* that makes them what they are. The existence of nitrogen as an element is inextricably woven into the situation when humans became able to fix more nitrogen than all nonhuman Earth processes combined.

WHAT CAN A CHEMICAL DO? NECROCHEMICALS, GREAT POWER/ GREAT RESPONSIBILITY, AND THE BECOMING CHEMICAL OF LIFE

When I asked one of the research group leaders of the pioneering Carbon Neutral Laboratory for Sustainable Chemistry at the University of Nottingham (one of the labs where I have done fieldwork as a resident Leverhulme fellow) what major breakthrough he would like to see in green and sustainable chemistry, he replied that we "need the equivalent of a Haber process for the twenty-first century that takes carbon dioxide and turns it into a benign product" (interview). Both of us are, of course, frightened by the effects that Haber-Bosch has had on the planet. But according to my interlocutor, the scale of CO_2 and its inextricable enmeshment with human life require a completely new vision and a new solution as deep and radical as the Haber-Bosch process. I have learned from CO_2 specialists that it is an ecologically contingent gas that lives not only in the actual worlds that we inhabit but also in many virtual worlds: it not only compels some humans to invent alternative sociotechnical ways of existing on a climate-changed planet but alters the human sense of planetary belonging and visions of ecological becoming. Elements and anthropogenic molecules shape human bodies and human action as much as human affects and imaginaries. More often than not, they shape our imaginaries of catastrophe and ecological destruction as we align ourselves with proliferating necrochemicals—such as the "ubiquitous elements" (Masco, this volume), including plutonium—that have shaped our sense of the future since the first test of a nuclear weapon in 1945.[3] Occasionally, however, future imaginaries entail a sense of collective obligation, that with the great power that elements give to humans comes great responsibility. Science fiction, movies, and comics have shaped much of these future imaginaries of collective social and ecological obligation.[4] Consider the conclusion of the film *Guardians of the Galaxy* (Gunn 2014), where the protagonists end war by sharing the burden of the elemental powers of the untamable and destructive purple infinity stone. In the film *Black Panther* (Coogler 2018), the peoples of the hidden and remote nation of Wakanda consider how to use the wealth and the powers that they have been able to harness from the element vibranium in the absence of colonial exploitation and white supremacy to benefit other marginalized peoples on the planet.

What is an element capable of? We do not know. To a certain extent, we know what some elements can do in certain ecologies. The ecologies in which chemicals emerge or are manufactured and their presence in them cannot be separated. And the chemical practices that make these molecules visible cannot be considered as separate from these events. Chemicals—both their ontic presence and our diverse knowledges of them—are bound together in their ecological context: the becoming chemical of life and the ecological becoming of chemical practice. Chemical substances are inessential.[5] They do not have proper boundaries; they change over time and they are contingent on the broader ecosystems they are part of. Different types of epistemic disciplinary regimes or different engineering industrial processes turn them into specific objects and set them in motion in the world. The chemospheres in which we exist are inhabited by such diverse substances. All these different epistemic, industrial, or other practices that make substances happen cannot be disentangled from the worlds in which they emerged and the worlds we live in.

When I talk about chemicals here, I refer to substances that are widely used or manufactured in human societies. I am focusing on anthropogenic chemicals that come to be visible through the diverse practices that take place in the ecologies in which they exist. Even the most abstract representation of chemicals in the standard nomenclature of chemical science are ecologically contingent, from the periodic table to the classification deployed for naming organic and inorganic compounds (see Hepler-Smith 2015; Klein 2013; Llored 2013). Chemistry is probably the most ecologically contingent science of all: it is a quintessentially empirical science that relies on the materialities of the labs and the extended spaces in which these labs live and on the relations of substances to other substances, apparatuses, social spaces, and human beings (see Bensaude-Vincent and Simon 2008; Hoffmann 1995; Lefèvre and Klein 2007). Anthropogenic chemicals are all deeply embedded and ecologically contingent forms of matter that exist simultaneously through multiple ecological registers.

Chemicals as mediators provide only a very limited view on the role substances play in intraspecies and interspecies exchanges. More than containing some essential properties that make mediation possible, what
is at stake in these substances is an ecological question: how these substances catalyze and organize relations within a whole nexus of organisms of one species, other species (including humans), and the broader
ecology. The multiplicity of chemicals—the chemodiversity within a certain ecology—is not just a functional means of communication but constitutes some of the very ontological conditions that sustain and modify
the life of different species within this specific context. When it comes to
toxins and semiochemicals, the ecology becomes continuous with plant
and animal bodies through chemical compounds. Bodies, compounds,
ecologies—all are made in the vortex of their interactions. Bodies make
worlds, worlds make chemicals, chemicals make relations between bodies, and so on (see Hustak and Myers 2013).

The identification of the molecular structure of pheromones gives
birth to something that is much wider than research on signaling relationships between plants and animals. It is the view of naturally occurring chemicals within an ecological context: pheromones are not just
signaling molecules; they are chemical practices of organisms that erect
their worlds. Ecology becomes part of chemical practice itself: the chemistry of biotic interaction as the organization of ecologies by molecular
means. The signaling relationships between plants and animals is not
just communication; these are *biochemical events* within the life of organisms in an ecology that is itself made through these interactions.

When such biochemical events are reduced to communication, they
are stripped by their processual-interactive qualities to become single,
operationalizable factors in the development of an organism. This is the
way by which ecologies are reduced to environments. The cobecoming
of ecology, organismic bodies, and chemicals is reduced to one single dimension that defines the context an organism inhabits. For example, in
epigenetic research in humans, the environment is usually reduced to
one single factor—such as nutrition, pollution, care, touch, conversation, social adversity, or stress—which, in turn, is conceived as a cue that
regulates the genome and suppresses or promotes gene transcription.

Although epigenetic research implicitly questions the reductionist view of gene activity as taking place solely within the gene and offers exciting new avenues for research, it often implies a different type of reductionism, which is about stripping the environment from its ecological nature (for an extended discussion, see Chung et al. 2016; see also Cromby et al. 2016; Landecker 2011; Papadopoulos 2011). Viewed from the perspective of the gene, anything else in the cellular environment beyond the DNA is considered environmental. At this level, how the environmental focus of the research can be conceptualized and operationalized can vary immensely. It is in this move that specific aspects of the wider ecology are often erased, while others are overemphasized and then isolated as single environments that cause these biochemical events.

Paradoxically, these different considerations of the environment do not necessarily assume that there is an ecological space to which they belong. Different aspects of the ecology are singled out as environments, and these environments are then condensed as triggers of biochemical events. Here we have environments without ecologies, and much of the environmental discourse today does exactly that: it refers to environments that are supposedly distinct from each other, independent in their workings and not ecologically contingent. Ultimately, environment is a human-centered concept that reduces the world to a bundle of certain features relevant to the specific processes under investigation.

THE AFTERLIVES OF CHEMICALS: INSURGENT ECOLOGIES, ECOLOGICAL CHEMISTRY, AND THE PERMANENCE OF POLLUTION

This is when ecologies become insurgent. They revolt against anyone or anything that will split them apart, single them out, and isolate them in order to control them. The 1960s saw not only the birth of biopesticides and the escalating war between humans and pests but also the improbable escape of many different ecologies from any attempt to tame them. The effects of anthropogenic chemicals were far more widespread, unpredictable, and complex: eutrophication, acidity, carcinogenesis, environmental persistence, global warming, endocrine disruption, extensive fossil fuel consumption, and so on.[6] In the moment when one believed one had found a way to control an environment, manufactured substances became uncontrollable themselves and started to spill over to other environments and to transform ecological organization. Environments do not exist alone; neither can they be seen as silos containing some few eclectic dimensions from the wider ecologies in which they belong. And yet it is this view that underlies much of mainstream chemical practice.

Tracing elements and toxicity in planetary circuits of pollution is a form of investigative chemical practice: ecological chemistry (see Joseph Masco, this volume). It is one of the disciplinary responses to ecological insurgency. The primary concern is to identify the fate of a chemical in ecological systems through the biomonitoring of environments, the construction of models of pollutants spreading in oceans and air and within soil, and the introduction of safe levels of exposure of biotic species to hazardous compounds (see Peijnenburg 2009). But the attempt to identify the effects of chemicals on ecologies necessitates solid infrastructures of environmental monitoring and reporting. And here is where the limits of ecological chemistry emerge: beyond scientific research, it requires the wide involvement of other actors, state agencies, governmental and intergovernmental institutions, industry, and private sector organizations that often do not cooperate or that even actively destabilize the process of monitoring. Reporting of environmental disasters and hazards is inaccurate and fragmented at best, while often it is intentionally misleading and erases the effects of chemicals (see, e.g., Brown and Mikkelsen 1990; Bullard 1994; Sheoin and Pearce 2014; Wylie 2017; see also Murphy, this volume). The extent of the damage and pollution is so widespread that the afterlives of chemicals when they are worlded tell us that surviving chemical contamination is not about protection. Sealing off is not possible.

There is no time left to waste in picking apart chemistry for the disasters its manufactured compounds have brought to Earth or in celebrating the extraordinary powers that chemicals have brought to humans. There are too many microbeads, toxins, solvents, and pollutants circulating through our bonded bodies, machines, and environments. Anthropogenic chemicals are so deeply entrenched within societies and ecologies that a life without chemical contamination is impossible to conceive of, let alone to realize. Imagining worlds without anthropogenic chemicals is impossible. There is no such thing as clean ecologies—not now, and not in the future. There is no way to step outside of the chemospheres we live in and exercise an external, wholesale critique of chemistry or, on the other hand, an indiscriminate glorification of chemicals. Instead, humble practical skills, molecular wisdom, and analytical patience can help to navigate the chemical landscapes that our bodies and ecologies have become. Someone once said that it is easier to imagine the end of capitalism than a world without anthropogenic chemicals.

INTO THE ZONE: LITTER IN HISTORY, OBLIGED BY ECOLOGY AND THE QUESTION OF REPARATION

In Andrey Tarkovsky's (1979) film *Stalker*, people enter a contaminated area known as the Zone to reach a mysterious wish-granting device. Although the Zone presents life-threatening dangers, people enter in order to fulfill their desires. Tarkovsky vehemently opposed a reading of the Zone as a metaphor or symbol for anything except life itself: a perilous, contaminated life in which people "may break down or come through," depending on their "capacity to distinguish between what matters and what is merely passing" (Tarkovsky 1989, 200). Yet all have to encounter the Zone and, if they survive, have to live in some way or another with its transformative effects. Whether or not they reach the device, they are changed by the Zone.

Arkady and Boris Strugatsky wrote the script for *Stalker* based on their science fiction novel *Roadside Picnic* (1977). In this and several other novels and short stories, the Strugatsky brothers explore the failure of humans to reconcile with their ecologies against the background of chemical environmental pollution in decaying postindustrial societies (see Maguire 2013). Litter is everywhere, and yet people are mesmerized by it. Each society creates its own litter and chemical waste. In fact, as in Dasgupta's (2009) novel *Solo*, we can read nineteenth- and twentieth-century history as a passage of different types of chemical production and contamination that correspond to specific periods of social organization. The more humans plunge into the litter they make, the more they distance themselves from their terrestrial ecologies and, paradoxically, the more they become entrenched in them. And the more they are unable to disentangle themselves from the decaying ecological spaces they create, the more they feel the "ecological call," which in Strugatsky's work takes the shape of an "ecological mysticism" (see Stableford 2005)—the sense of being obliged by ecology.

It is here that the epistemic register of ecology (as the branch of biological science) fuses again—even if only provisionally—with the ontic, experiential, and (eventually) political register. Humans producing chemicals have an obligation for ecological accountability, which starts with the investigation of the afterlives of manufactured chemicals. And then there is an ethical requirement for a response. Ecology as the underlying "philosophy" of ecological movements (see Hay 2002) and ecologically conscious scientists (see, e.g., Gorman 2013) inserts itself into chemical practice as the effects of anthropogenic chemicals become the target of ecological and social justice movements.[7] Chemical practice becomes obliged by ecology.

Becoming obliged involves many and often very disparate affective, intellectual, and practical dimensions (for a detailed discussion of obligation, see Puig de la Bellacasa 2015; 2017, 150ff; this volume). There is, of course, a sense of obligation toward the urgency of ecological questions of interconnectedness and sustainability. It is about being compelled, feeling a requirement and, perhaps, the duty to respond to current ecological crises. It also involves some sense of feeling grateful, indebted to the ecological wonders of the nonhuman world. Obligation often comes as a response to the campaigns of ecological movements. Here it is associated with a commitment to unsettle militarized and colonial ecologies. Puig de la Bellacasa (2017) highlights that obligation is not only about a sense of moral responsibility but involves a practical-material dimension and the urgency for action: being obliged by ecology means that one has no choice but to care for the ecologies one is part of or to face destruction. Obligation gravitates around the quest to "repay" for damages done and to launch ways of reparation. Within chemical ecology, obligation and reparation are inseparable. Reparation—both as repair and as compensation for and restoration of the damage done—emphasizes this material dimension of being obliged by ecology.

So, beyond the commitment to ecology as a biological science and the obligation for ecological accountability within ecological chemistry, in what follows I will sketch two configurations of ecological obligation that engage with the sense of reparation: first, green chemistry that, in principle, attempts to create molecules that incorporate ecological contingency in the process of their making and the molecules themselves, and second, autonomous generative chemical experimentation in the interstices of instituted and community research and practice that is driven by questions of ecological care.

When chemist Barry Trost (1991) coined the term *atom economy*, he meant the reduction of atom waste in chemical reactions. Reactions ideally should incorporate all the atoms of the reactants. This is a minimal step but one that has had a widespread impact in imagining chemical research in ways that attempt to incorporate ecological concerns into the reaction itself. Almost simultaneously, Roger Sheldon (1992, 2007) introduced a very simple, powerful, and effective way to approach the problem of chemical waste: "I called it the 'E factor,' the environmental factor, so that a number of kilograms of waste per kilogram of product" (Interview). From today's perspective, it is inconceivable how such a plausible approach to waste as the one created by Roger Sheldon evaded chemical research, although it was such an obvious and pressing question for decades. And indeed, there is no clear explanation for this terrifying delay in human capacity to understand the wastefulness of our earthly existence. Roger Sheldon's foundational research was a turning point toward an ecologically obliged chemical practice, which expanded later to account for rethinking chemistry holistically and for contributing to the emergence of "green and sustainable chemistry."

The aim was to produce research on waste-free, nontoxic, low-impact compounds. The end of the 1980s saw the first ideas on green chemistry emerging, and the 1990s saw its establishment as an alternative pathway to standard chemical innovation (see Iles and Mulvihill 2012; Linthorst 2010; Logar 2011; Matus 2010; Jody Alan Roberts 2005; Woodhouse and Breyman 2005), Driven by the aim to be "benign by design" rather than an attempt to limit, regulate, and manage the handling and effects of chemicals, green chemical practice comes as a realization of ecological chemistry's approach that of all chemical products and processes in existence, perhaps only 10 percent are already environmentally benign. For example, if you look at the full life cycle of a traditional drug, on average, "for every kilogram of pharmaceutical drug that you're generating, you're generating a tonne of hazardous" waste (Sanderson 2011, 19).

Around the mid-1990s, a series of prominent initiatives helped establish the field, such as the creation of the US Environmental Protection Agency's (EPA) Green Chemistry Program.[8] The year 1998 saw the launch of the Green Chemistry Network of the Royal Society of Chemistry in Britain and the incorporation of the Green Chemistry Institute, which very soon after became part of the American Chemical Society. Simultaneously there was the influential and now classic publication of *Green Chemistry: Theory and Practice* by Paul Anastas and John Warner (1998) and one year later the publication of the journal *Green Chemistry* by the Royal Society of Chemistry (Anastas et al. 2016; Clark et al. 2014).

HOLISTIC CHEMICAL PRACTICE: LIFE CYCLE ANALYSIS, NO LIKE FOR LIKE, AND THE F-FACTOR

Green chemistry aims to be the molecular basis of sustainability. While atom economy and the broader attempt of E-Factor research targeting the reduction of chemical and other waste were at the core of the beginning of green chemistry, there is a concern for a more holistic approach that encompasses resources and the sourcing of feedstocks, energy expenditure and efficiency, and the afterlife of chemicals and products (see Anastas and Zimmerman 2019; Clark et al. 2014; Poliakoff, Licence, and George 2018). Bringing together waste, resources, and postproduction life means to complete the production cycle of a molecule. This form of circular chemical economy mobilizes ecology as the totality of the relations of a molecule. Ideally, green chemistry attempts to look at the full life cycle of a chemical.

Marking this attempt as an ideal constellation is important, because embracing an ecological life-cycle perspective has far-reaching consequences and troubles widespread assumptions about the role green chemistry can play in rethinking chemical research as a whole (for various approaches, see Borrion et al. 2018; Caillol 2013; Lokesh, Ladu, and Summerton 2018; McManus and Taylor 2015). Broadly understood, green chemistry is expected to substitute existing products with "green" ones. Consider, for example, the most common plastic used worldwide—polyethylene (PE)—with a global demand of 99.6 Mtonnes in 2018 (Fortune Business Insights 2020; Freedonia Group n.d.). One approach to "greening" PE would be to replace the petrobased feedstock with one made from renewable raw materials. Green PE, such as I'm Green™ Polyethylene (Braskem USA n.d.), is produced from sugarcane ethanol in a process that reduces waste, reduces energy consumption, and delivers a product with very similar properties to petrochemical polyethylene that can be recycled in the same way. However, from the perspective of green chemistry, that would not be enough.

The same material but with some green attributes is not a strong criterion for green chemistry, especially when it involves a full life-cycle analysis.[9] The problem is quite simple: green PE becomes an actual object, such as a bottle, which—like one made from petrobased PE—is not compostable and not biodegradable. The bioaccumulation of green PE is very high, and because it is so widely used, recycling is very limited compared to the overall amount of production. In addition, a full life-cycle approach would involve looking at the sourcing of feedstocks from renewable materials. The replacement of food crops with crops used for chemical production, which has been widely discussed, raises concerns about working conditions in the sugarcane fields, the destructive effects of sugarcane monocultures on soils, the unequal geopolitics of chemical feedstocks, extractivism in the Global South, and other issues. A full life-cycle approach deeply troubles what James Clark, director of the innovative Green Chemistry Centre of Excellence at the University of York, calls our expectation for "like for like as a replacement" (Interview). Questioning the possibility of "like for like" has tremendous implications not only for what we consider to be green chemical practice but also for what it can and cannot do in today's socioeconomic and technoscientific conditions. It becomes apparent that we need a deep understanding of what one of the leading green chemists in the UK, Martyn Poliakoff (2014, 21), calls the F-Factor (*F* for *function*), which "would enable people to compare the amounts and also the composition of different chemicals needed to provide a particular function" (Interview). "We have to make every atom count," as Pete Licence, director of the pioneering Carbon Neutral Laboratory at the University of Nottingham, says. "So, we make a molecule that delivers a function, rather than delivers the function by accident, which is often the way that chemistry has worked. Because, we might make a molecule and that molecule is elegant, but we then look for applications for that molecule, but what we're trying to do is to think about the application and design the molecule in the most environmentally appropriate and most efficient way" (Interview).

VALUATION CLASHES: MOLECULAR VALUE, ECOLOGICAL VALUE, AND THE LIMITS OF GREEN

So, what would an ecologically minded green chemistry replace? The feedstock, with or without its sourcing? Parts, or all, of the chemical's production process? All of the above, including the molecule itself? Or all aspects of the creation of the molecule, including its potential applications and afterlives? Depending on the answer, different forms of green chemical practice are possible. But it is apparent that the more we expand chemical practice to encompass larger aspects of the life cycle of a molecule, the more we reach the limits of green and sustainable chemical practice. The more holistic it is, the more difficult chemical practice is to adopt, because it challenges existing research, production process infrastructures, policy and regulation, consumer habits, and so forth. In other words, the more holistic, or ecological, chemical practice becomes, the less stable is its molecular, or market, value. There is a clash of valuation here (Lilley and Papadopoulos 2014), a clash between ecological value and molecular/market value. In current sociomaterial and economic conditions, these two scales of value cannot be fully brought together without the reduction of one to the other. There are possibilities for moving one slightly toward the other, but fully merging them seems almost impossible.

Even if we were able to produce fully green compounds of today's ten most used chemicals (and we are not), within the current socioeconomic and cultural conditions we would not be able to use them, simply because there are so many barriers to adoption that would not permit such a change. Chemistry *itself* is a socioecologically contingent science, and this sets the barriers for a green and sustainable chemistry. Green chemistry is an urgent undertaking for sustaining livable words. There is no future without green chemistry and yet realizing green chemistry in its ecologically holistic form is very difficult.

This is why green chemistry appears always as a compromise that attempts to shift the balance from molecular value on the one hand toward ecological value on the other. But a total shift toward green molecules is not possible in today's conditions, where molecular value is required for a compound to be produced and used widely. A full-scale green chemistry is an absolute necessity, and simultaneously it is impossible as a full-scale process. Green and sustainable chemistry needs a long time to be developed and adopted, but the current pace of ecological damage is much faster. This does not diminish in any possible way the importance and urgency of green and sustainable chemistry; rather, it reveals its ecological limits. To paraphrase the title of Roy Lichtenstein's 1970 lithograph and silkscreen print *Peace through Chemistry I*, one could say that if there will not be peace through *green* chemistry, then there will not be peace through chemistry at all. But the importance and urgency of green chemistry confront the question of the temporal scale of molecules themselves.

THE PROBLEMATIC OF SCALE: THE INFERNAL BINARY OF GROWTH/ NO GROWTH AND THE ECOCOMMONING OF PLANETARY BOUNDARIES

The impossibility and urgency of green chemistry are exemplified in the problem of scale. A chemical is said to be "real" or to "have legs" when it can be manufactured on a multikilogram or tonne scale, depending on the type of compound. Within science and technology studies, there have been various problematizations of scale. The approaches vary significantly, of course. But many of them approach scale as problematic—in fact, as something that might not even exist as such. For example, Strathern (2000), Jensen (2015), and others make the case that rather than nesting scales, we have fractal environments whose complexity is scale invariant, as they incorporate many different qualities that traditionally belong to different scales. Tsing (2012, 2015) goes a step further and elevates scalability to a device that performs domination and ultimately social and ecological destruction.

However, petrobased, toxic, and hazardous chemicals are produced and used at such scale that we need a tremendous economy of scale of alternative chemicals in order to replace them. Scale is ambivalent. Rather than one single core thing, scale becomes a problematic: it is the organizing principle of ecological degradation, and simultaneously it is a necessary component of many reparation and remediation attempts. Without some form of scale, there is no possibility to confront the extent to which anthropogenic chemicals pervade everything. However, as we know it, scale is attached to growth. And growth is attached to ecological destruction. But the problematic of scale complicates our relation to growth. This is because the scale of toxic chemicals and the scale of alternative benign chemicals that are needed to reduce further ecological degradation sit uneasily with both the celebration of growth and its outright dismissal. Limitless growth and no growth, fast and slow, scale and small become impossible to be decided in practical terms; they are "infernal alternatives" (Pignarre and Stengers 2011, 24). The category of growth is too universal to be encountered. Its absence mirrors the universality of its presence. If its presence is associated with ecological degradation and loss of nonhuman life, its absence is imposed on those who already feel its destructive consequences: disadvantaged and marginalized social groups, the Global South, large ecosystems, and other species. Steady and pure growth has been always a historical and geopolitical anomaly reserved for small parts of the Global North.

Growth aside, what do we do with the planetary and social boundaries to growth that we are currently facing?[10] This is where the necessity for a different practice of scale arises—neither as an answer to growth nor as an alternative to it, but because the presence of growth has made any meaningful engagement with social and planetary boundaries impossible. Rather than adopting impossible alternatives, a multiplicity of ecological movements create conditions for establishing their own boundaries and creating their own scales (see Kallis 2019). The process of constructing these adequate scales is a process of commoning our boundaries, or ecocommoning—the making and use of communally maintained spaces and ecologies that humans partake of (see Papadopoulos 2018, especially chapters 5 and 8; see also Bresnihan 2013; Castellano 2017; Linebaugh 2008, 2010; Reid and Taylor 2010; Wall 2014).

INTENSIVE SCALES: FROM EXTENSIVE MEASUREMENT
TO SINGULAR INVOLVEMENTS WITH CHEMICALS

The scale of a chemical is the ecology of the chemical practice that engages with it (see Stengers, this volume).[11] As substances move to different scales, they also become different objects: their ontological existence and their relations to other substances change.[12] But scaling up, and occasionally scaling down, present enormous technical and engineering challenges. Many of the green chemists I talked to mentioned that one of their dreams and a definite measure of academic success would be if the molecules they are working on can be scaled up, possibly and ideally all the way to becoming readily available on a multikilogram or even an Mtonne scale to be used by consumers. Moving from a milligram or gram level to kilogram and then to a tonne level is inextricably mixed with ecological concerns. In fact, a molecule is not the same on a different scale—that is, in a different ecological context (see Hayden's [2007, 2010, 2011] important work on sameness and difference of chemical compounds). A scale is not a quantitative index of the mass or volume of a molecule. It is an ecology that incorporates different material, technological, sociocultural, and temporal constraints that define what a molecule is as it moves between bodies and through worlds.

Thus, a different understanding of scale emerges here: instead of understanding scale as a device that captures incremental changes—that is, scale as extensive measurement—I approach scale as transformative processes situated and dependent on their locations and the ecologies they are embedded in. Intensive scales negotiate life within toxic regimes by involving substances in ways that trigger minor qualitative changes in their ecologies. Rather than extension through replication or through the increase of volume or mass, intensive scales are about the creation of an abundance of many different ecologically contingent chemical practices.

Intensive scales imply that molecules do different things in different ecologies. A PET (polyethylene terephthalate) bottle or an HDPE (high-density polyethylene) milk bottle, for example, can be engaged in radically different ways, depending on the context. During my work with the Leicester Hackspace, UK (http://leicesterhackspace.org.uk/), I encountered several people using PET and HDPE to experiment with different processes of recycling and reuse. For one of these hackers, Woody Kitson, the process of repurposing involves "the reprocessing (e.g., melting, reforming, welding, cutting, sewing, turning, milling) of the raw material" which then can be used in many different ways: as a container; as a nonconductive spacer; for toys, tools, or tent pegs; as moldable plastic, if treated correctly; as replacement parts for broken plastic components; as material for art projects; as one of the ingredients of composite products; as material for experimentation to learn about the toxic chemicals and other hazardous ingredients it possibly contains; as a source of durable string and so forth (see Papadopoulos 2018).

The link between the Leicester Hackspace, multiple human and nonhuman actors within the space, and polyethylene created a scale of experimentation that is radically different from standard research, manufacturing, or consumer settings where this material is widely used. The effects of this engagement are radically different too: citizen sharing of knowledge, community technoscience, the raising of environmental consciousness, the development of educational programs about waste, toxicity, and reuse, the development of grassroots recycling initiatives, and so on. And while the engagement with polyethylene within the Leicester Hackspace might be considered to be very limited in its scale and scope, it is similar to other experimentations with materials that I encountered in other fablabs and hacker and maker spaces. Scale here is not about increase of mass, volume, or monetary value but about the multiplication of ecologically contingent ways to engage with a problematic—that is, the toxic overabundance of plastics in everyday life. What if we approach scale not as a question of extension but as an intensive experimental field in which chemical practice is distributed across different ontologies, practices, and knowledge systems?[13] What if we see scale as the multiplication of experimenting with chemical substances in different locales that create singular transformative involvements with these chemicals? Could chemical practice thereby become ecological?

GENERATIVE CHEMICAL PRACTICE: AUTONOMOUS CHEMISTRIES AND REPARATIVE JUSTICE

Rather than thinking about how life could once again become free of toxic chemicals, the question is how to live in chemically contaminated worlds (see Liboiron, Tironi, and Calvillo 2018)—how to develop a chemical practice that sustains life in toxic worlds and creates paths of regeneration (see, e.g., R. Lee 2020). There is no global and final fix for our chemospheres, and certainly there is no possibility to step outside them. What remains is to create many small paths of healing. Such paths are generative; they involve direct and contingent experimentation with the singular conditions in which chemical substances emerge and the specific characteristics of these substances in order to combat toxicity and to create alternative healing compounds. I call generative chemical practice these experimental paths that create many different minor worlds of reparation.

Such a chemical practice is autonomous of human needs (see Lotringer and Marazzi 1980; for a discussion of the concept of autonomy, see Böhm, Dinerstein, and Spicer 2010; Papadopoulos 2018; Papadopoulos, Stephenson, and Tsianos 2008). It is not disconnected from them, but it takes distance from human predominance in order to cultivate wider interdependence. Its primary commitment is toward its ecological grounding. Autonomous generative chemical practice is an act of ecological self-care. In its core is the commitment to what binds things together, an attention to the wholeness of the practice itself. Rather than thinking of chemical practice as one single scientific endeavor, it can be seen as a multiplicity of many different ontological acts of healing. A generative chemistry is possible; a generative chemistry is here.

Chemical practice becomes dispersed into the "distributed invention power" (Papadopoulos 2018, 182) of amateur scientists, Indigenous knowledge practitioners, clandestine chemists, do-it-yourself biochemists, remediation ecologists, biodegradable designs, underground labs, and interspecies collaborations. It becomes an integral part of community technoscience, in the developing of entheogens, healing compounds, ethnobotanical knowledge, and kitchen chemistries; in the baking of bread, the making of beer, the mattering of compost; in the development of amateur-led pollution-sensing devices and the monitoring of chemical toxicity; in the creation of vocabularies, images, and stories to capture life in contaminated worlds; in independent chemical experimental lab work; and in the incorporation of green chemistry within citizen science. Amateur chemical practitioners create alternative ontologies, alterontologies, on the molecular level. Ecology is not just embedded in the making of chemical innovation, it is its very condition. Generative chemical practice is not only obliged to ecological science, ecological thinking, and ecological movements, but it is existentially obliged to the very ecology it is embedded in.

Yet generative chemical practice is neither a break from green and sustainable chemistry, nor does it entail a separation from the epistemic worlds of other forms of chemical practice. Rather, generative chemical practice is their continuation, which fully reinserts chemical practice into ecology—in multiple registers—to encounter the challenge of reparation. A split between the epistemic frame of anthropogenic chemicals, the engineering frame of industrial production, and the ontic frame of substances that make up our worlds is neither existentially possible nor politically desirable. Generative chemical practice is not oppositional to other forms of chemical practice; rather, it is directly connected to them by absorbing what it needs from them in order to practice elemental reparation.

Generative chemical practice is not about segregation. Reparation is not about separation. There is no rupture here but a departure. There is no opposition but a withdrawal for reductive forms of engaging with chemicals. Generative chemical practice is about becoming elemental (see, e.g., Cohen and Duckert 2015; Macauley 2010; Peters 2015). Rather than engaging solely with single elements in reduced environments, generative chemical practice involves anthropogenic chemicals in larger, nonreductive ecologies, other than human lives and a multiplicity of human stories—epistemic and vernacular, Indigenous and nomadic, impermanent and profound. Generative chemical practice is hybridizing both the elements and the elemental by remixing the elements of the periodic table, our synthetic molecules, and our anthropogenic compounds with the elemental worlds, which are sustained or consumed by water, fire, air, and earth.[14]

Striving for reparation of contaminated worlds is a quest for justice. Reparative justice is simultaneously about making amends owed for wrongful harm *and* about repairing damaged ecologies. It is about the minor healings of generative chemical practices and the decolonization of our ecologies of existence (see Ferdinand 2019; Murphy 2017; Shadaan and Murphy 2020). Reparative justice is disanthopocentric and attends to the conditions of the human and nonhuman (the elemental and the ecological) rather than only to what makes up the human and nonhuman (separate elements and substances). It involves chemical cooperation and the making of chemical infrastructures obliged by ecology. Reparative justice involves holistic life-cycle practices, an intensive ecological approach to scale, and distributed invention power. All these are different paths that strengthen the generative dimension of chemical practice as ecological reparation. But at its core, ecological reparation asks a fundamental question of justice: Who has the freedom to decide what chemical substances to create and at what cost, and how is this freedom achieved?

ACKNOWLEDGMENTS

I gratefully acknowledge support from the Leverhulme Trust, UK (RF-2018-338\4). As always, my debt of gratitude is owed to the scientists and researchers who responded to my requests for conducting fieldwork in their labs and for sharing their insights into green and sustainable chemical practices.

NOTES

1. Ecological science itself is embedded in the wider social, environmental, and historical settings of colonial expansion, travel writing, and assuming the world from the explorers' perspective (Ferdinand 2019; Gómez-Barris 2017; Pratt 1992). The ecological sciences were invented, at least in North America, as part of a larger socioepistemic movement that facilitated colonization and dispossession of native lands (see, e.g., Kingsland 2005).

2. As Wylie (2012, 63) points out, "By 1952, DDT resistance 'had been developed by important pests of apples, cabbages, potatoes, tomatoes and grapes [as well

as] the body louse, the bedbug, two species of fleas and several species of mosquitoes' (Brown, 1977, 22). Furthermore, resistance to the second round of synthetic chemical pesticides, including Aldrin and Malathion, was also increasing. Aldrin, part of the 'cyclodiene group of organochlorines,' was introduced in 1948, and resistant pests were already emerging by 1955, three years before this advertisement (Brown, 1977, 23). Meanwhile resistance to organophosphorus pesticides like Malathion had been seen as early as 1949 (Brown, 1977, 24)."

3. Such a sense of alignment with the "evil" is exemplified in the account of the scientist in charge of the first nuclear bomb test detonation (Bainbridge 1975, 46).

4. The theme of power and responsibility runs through many cultural devices in the Marvel universe over a period of more than fifty years, from Spiderman's discovery of his sense of duty (Lee and Ditko 1962, 10) to *Black Panther* (Coogler 2018). See also Hecht 2012. For a list of fictional elements, see https://en.wikipedia.org/wiki/List_of_fictional_elements,_materials,_isotopes_and_subatomic_particles.

5. Materials, molecules, and substances are not essential entities with fixed properties. As Whitehead (1925, 111) says, they are "continuously inheriting a certain identity of character transmitted throughout a historical route of events. This character belongs to the whole route, and to every event of the route. This is the exact property of material." Barry (2005, 52) adds that "molecules should not be viewed as discrete objects, but as constituted in their relations to complex informational and material environments."

6. Anthropogenic chemicals are powerful substances that change ecological organization. These chemicals include plastics (Jody A. Roberts 2010) and microparticles (Eriksen et al. 2014); endocrine-disrupting chemicals (EDCs) such as bisphenol A (BPA), phthalates, herbicides, and pesticides (Gore et al. 2015); anthropogenic volatile organic compounds (VOCs) such as fossil fuels, most of the solvents used in paints, benzene, formaldehyde, and chlorofluorocarbons (Shapiro 2015); and persistent organic pollutants (POPs) such as polychlorinated biphenyls (PCBs), dioxins, and several synthetic chemical pesticides (e.g., dichlorodiphenyltrichloroethane (DDT) and aldrin.

7. Beginning in the 1940s, the influence of ecology on shaping environmental movements has been profound—particularly seen in such classic texts as Hutchinson et al. 2010; Odum 1953.

8. This development was followed by a US Presidential Green Chemistry Award competition, which was formally announced in March 1995, and a series of conferences on green chemistry run by the National Academy of Science and the National Academy of Engineering.

9. I am inspired by the discussion of sameness and difference of chemical substances in Hayden (2007, 2012). See also Hoffmann 1995.

10. There are many different conceptualizations of planetary boundaries. The most widespread (and probably the most universalist) can probably be found in Rockström et al. (2009). See also Diamond et al. 2015.

11. "This is how I produced what I would call my first step towards an ecology of practice, the demand that no practice be defined as 'like any other,' just as no living species is like any other. Approaching a practice then means approaching it as it diverges, that is, feeling its borders, experimenting with the questions which practitioners may accept as relevant, even if they are not their own questions, rather than posing insulting questions that would lead them to mobilise and transform the border into a defence against their outside" (Stengers 2005, 184).

12. Dumit (this volume) proposes a method of following the substance to figure out the various scales and locations that it assembles around it.

13. I am using the concept of the intensive as developed by Deleuze and Guattari (1987), especially in chapter 2, and as summarized by Bonta and Protevi (2004): "One can call 'intensive' any linked set of rates of change in assemblages or 'rhizomatic multiplicities' since changes in these ratios past an immanently determined critical threshold also trigger qualitative change. . . . The key to Deleuze's ontology is the claim that intensive morphogenetic processes give rise to actual or stratified entities whose extensive properties and fixed qualities are the object of representational thought and occlude the intensities which gave rise to them" (101).

14. Paracelsus (1996), who is much better known for his pharmacology and his contribution to toxicology, believed in the existence of sentient beings, elemental spirits that are made of and live within the elements of earth, water, air, and fire. These liminal, sensible beings—gnomes, undines, sylphs, and salamanders—carry the material knowledge and sensibilities of each element. These liminal supporters of Gaia inhabit worlds that are not possible for humans to relate to directly, says Paracelsus. But there are paths of sensing and acting that allow humans to experience their presence and their commitment to Earth. Within these protoecological European imaginaries, the becoming elemental of chemical practice would involve people learning to move through the soils as gnomes, chthonic beings, the beings of the Earth that work with physical matter to create lasting environments. Gnomes are the heart of earthly households. Some humans will become united with water as undines, the ludic elementals who engage with movements and experience the force of waves and water currents and flows. Some humans will embody air as sylphs, beings united with the skies and the wind, who move lightly and invisibly through air, connecting flowers and trees together and making community. Sylphs are air; sylphs are our breath, and humans can become their own breath too. "Conspire!" as Tim Choy says in his contribution to this volume. Salamanders, the beings of fire, show power, intensity, ardor; they purge with fire. Certain humans can embody these energies which, although very stirring and difficult to control, are indispensable for cleansing, healing, and regeneration.

REFERENCES

Anastas, Paul T., Buxing Han, Walter Leitner, and Martyn Poliakoff. 2016. "'Happy Silver Anniversary': Green Chemistry at 25." *Green Chemistry* 18 (1): 12–13. https://doi.org/10.1039/c5gc90067k.

Anastas, Paul T., and John Charles Warner. 1998. *Green Chemistry: Theory and Practice*. New York: Oxford University Press.

Anastas, Paul T., and Julie B. Zimmerman. 2019. "The Periodic Table of the Elements of Green and Sustainable Chemistry." *Green Chemistry* 21 (24): 6545–66. https://doi.org/10.1039/c9gc01293a.

Bainbridge, Kenneth T. 1975. "A Foul and Awesome Display." *Bulletin of the Atomic Scientists* 31 (5): 40–46.

Baker, Raymond, Richard Herbert, Philip E. Howse, Owen T. Jones, Wittko Francke, and Wolfgang Reith. 1980. "Identification and Synthesis of the Major Sex Pheromone of the Olive Fly (Dacus oleae)." *Journal of the Chemical Society, Chemical Communications* 2:52–53.

Barry, Andrew. 2005. "Pharmaceutical Matters: The Invention of Informed Materials." *Theory, Culture and Society* 22 (1): 51–69. https://doi.org/10.1177/0263276405048433.

Bensaude-Vincent, Bernadette, and Jonathan Simon. 2008. *Chemistry: The Impure Science*. London: Imperial College Press.

Bensaude-Vincent, Bernadette, and Isabelle Stengers. 1996. *A History of Chemistry*. Cambridge, MA: Harvard University Press.

Böhm, Steffen, Ana C. Dinerstein, and André Spicer. 2010. "(Im)possibilities of Autonomy: Social Movements in and beyond Capital, the State and Development." *Social Movement Studies* 9 (1): 17–32. https://doi.org/10.1080/14742830903442485.

Bonta, Mark, and John Protevi. 2004. *Deleuze and Geophilosophy: A Guide and Glossary*. Edinburgh: Edinburgh University Press.

Borrion, Aiduan, Jun Matsushita, Kat Austen, Charlotte Johnson, and Sarah Bell. 2018. "Development of LCA Calculator to Support Community Infrastructure Co-design." *International Journal of Life Cycle Assessment* 24 (7): 1209–21. https://doi.org/10.1007/s11367-018-1492-2.

Braskem USA. n.d. "I'm Green." Accessed March 15, 2021. https://www.braskem.com.br/imgreen/home-en.

Bresnihan, Patrick. 2013. "John Clare and the Manifold Commons." *Environmental Humanities* 3:71–91.

Brown, Phil, and Edwin J. Mikkelsen. 1990. *No Safe Place: Toxic Waste, Leukemia, and Community Action*. Berkeley: University of California Press.

Bullard, Robert D. 1994. *Dumping in Dixie: Race, Class, and Environmental Quality*. Boulder, CO: Westview.

Butenandt, A., R. Beckmann, D. Stamm, and E. Hecker. 1959. "Über den Sexual-Lockstoff des Seidenspinners Bombyx mori: Reindarstellung und Konstitution." *Zeitschrift für Naturforschung B: A Journal of Chemical Sciences* 14 (4): 283–84.

Caillol, Sylvain. 2013. "Life Cycle Assessment and Ecodesign: Innovation Tools for a Sustainable and Industrial Chemistry." In *The Philosophy of Chemistry: Practices, Methodologies, and Concepts*, edited by J. Llored, 35–64. Newcastle upon Tyne, UK: Cambridge Scholars Publishing.

Castellano, Katey. 2017. "Moles, Molehills, and Common Right in John Clare's Poetry." *Studies in Romanticism* 56 (2): 157–76.

Chung, Emma, John Cromby, Dimitris Papadopoulos, and Cristina Tufarelli. 2016. "Social Epigenetics: A Science of Social Science?" *Sociological Review* 64 (1): 168–85. https://doi.org/10.1111/2059-7932.12019.

Clark, James H., Roger Sheldon, Colin Raston, Martyn Poliakoff, and Walter Leitner. 2014. "15 Years of Green Chemistry." *Green Chemistry* 16:18–23.

Cohen, Jeffrey Jerome, and Lowell Duckert. 2015. *Elemental Ecocriticism: Thinking with Earth, Air, Water, and Fire*. Minneapolis: University of Minnesota Press.

Coogler, Ryan, dir. 2018. *Black Panther*. Burbank, CA: Marvel Studios.

Cromby, John, Emma Chung, Dimitris Papadopoulos, and Chris Talbot. 2016. "Reviewing the Epigenetics of Schizophrenia." *Journal of Mental Health* 25 (6): 1–9. https://doi.org/10.1080/09638237.2016.1207229.

Dasgupta, Rana. 2009. *Solo*. London: Fourth Estate.

Deleuze, Gilles, and Félix Guattari. 1987. *A Thousand Plateaus: Capitalism and Schizophrenia*. Minneapolis: University of Minnesota Press.

Diamond, M. L., C. A. de Wit, S. Molander, M. Scheringer, T. Backhaus, R. Lohmann, R. Arvidsson, A. Bergman, M. Hauschild, I. Holoubek, L. Persson, N. Suzuki, M. Vighi, and C. Zetzsch. 2015. "Exploring the Planetary Boundary for Chemical Pollution." *Environment International* 78:8–15. https://doi.org/10.1016/j.envint.2015.02.001.

Dillon, Lindsey. 2014. "Race, Waste, and Space: Brownfield Redevelopment and Environmental Justice at the Hunters Point Shipyard." *Antipode* 46 (5): 1205–21. https://doi.org/10.1111/anti.12009.

Eriksen, M., L. C. Lebreton, H. S. Carson, M. Thiel, C. J. Moore, J. C. Borerro, F. Galgani, P. G. Ryan, and J. Reisser. 2014. "Plastic Pollution in the World's Oceans: More than 5 Trillion Plastic Pieces Weighing over 250,000 Tons Afloat at Sea." *PLoS One* 9 (12): 1–15. https://doi.org/10.1371/journal.pone.0111913.

Fabre, Jean-Henri. 1912. *Social Life in the Insect World*. London: Unwin.

Ferdinand, Malcom. 2019. *Une écologie décoloniale: Penser l'écologie depuis le monde caribéen*. Paris: Éditions du Seuil.

Fortun, Kim. 2014. "From Latour to Late Industrialism." *HAU: Journal of Ethnographic Theory* 4 (1): 309–29. https://doi.org/10.14318/hau4.1.017.

Fortune Business Insights. 2020. "Polyethylene Market Size, Share and COVID-19 Impact Analysis." September. https://www.fortunebusinessinsights.com/industry-reports/polyethylene-pe-market-101584.

Freedonia Group. n.d. "World Polyethylene." Accessed January 14, 2020. http://www.freedoniagroup.com/industry-study/world-polyethylene-3210.htm.

Gómez-Barris, Macarena. 2017. *The Extractive Zone: Social Ecologies and Decolonial Perspectives*. Durham, NC: Duke University Press.

Gore, A. C., V. A. Chappell, S. E. Fenton, J. A. Flaws, A. Nadal, G. S. Prins, J. Top-
pari, and R. T. Zoeller. 2015. "Executive Summary to EDC-2: The Endocrine
Society's Second Scientific Statement on Endocrine-Disrupting Chemicals."
Endocrine Reviews 36 (6): 593–602. http://www.ncbi.nlm.nih.gov/pubmed
/26414233.

Gorman, Hugh S. 2013. *The Story of N: A Social History of the Nitrogen Cycle and the
Challenge of Sustainability*. New Brunswick, NJ: Rutgers University Press.

Guattari, Félix. 1995. *Chaosmosis: An Ethicoaesthetic Paradigm*. Bloomington:
Indiana University Press.

Gunn, James, dir. 2014. *Guardians of the Galaxy*. Burbank, CA: Marvel Studios.

Haniotakis, George E., W. Francke, K. Mori, H. Redlich, and V. Schurig. 1986.
"Sex-Specific Activity of (R)-(–)- and (S)-(+)-1,7-dioxaspiro[5.5]undecane, the
Major Pheromone of Dacus oleae." *Journal of Chemical Ecology* 12 (6): 1559–68.

Hay, P. R. 2002. *Main Currents in Western Environmental Thought*. Bloomington:
Indiana University Press.

Hayden, Cori. 2007. "A Generic Solution? Pharmaceuticals and the Politics of the
Similar in Mexico." *Current Anthropology* 48 (4): 475–95.

Hayden, Cori. 2010. "The Proper Copy: The Insides and Outsides of Domains
Made Public." *Journal of Cultural Economy* 3 (1): 85–102. https://doi.org
/10.1080/17530351003617602.

Hayden, Cori. 2011. "No Patent, No Generic: Pharmaceutical Access and the Politics of the Copy." In *Making and Unmaking Intellectual Property: Creative Production in Legal and Cultural Perspective*, edited by Mario Biagioli, Peter Jaszi,
and Martha Woodmansee, 285–304. Chicago: University of Chicago Press.

Hayden, Cori. 2012. "Population." In *Inventive Methods. The Happening of the Social*,
edited by Celia Lury and Nina Wakeford, 173–84. London: Routledge.

Hecht, Gabrielle. 2012. *Being Nuclear: Africans and the Global Uranium Trade*. Cambridge, MA: MIT Press.

Hepler-Smith, Evan. 2015. "'Just as the Structural Formula Does': Names, Diagrams, and the Structure of Organic Chemistry at the 1892 Geneva Nomenclature Congress." *Ambix* 62 (1): 1–28. https://doi.org/10.1179/17458234
14y.0000000006.

Hoerl, Erich. 2013. "A Thousand Ecologies: The Process of Cyberneticization and
General Ecology." In *The Whole Earth: California and the Disappearance of the
Outside*, edited by Diedrich Diederichsen and Anselm Franke, 121–30. Berlin:
Sternberg.

Hoffmann, Roald. 1995. *The Same and Not the Same*. New York: Columbia University Press.

Hustak, C., and Natasha Myers. 2013. "Involutionary Momentum: Affective Ecologies and the Sciences of Plant/Insect Encounters." *differences* 23 (3): 74–118.
https://doi.org/10.1215/10407391-1892907.

Hutchinson, G. Evelyn, David K. Skelly, David M. Post, Melinda D. Smith, and
Thomas E. Lovejoy. 2010. *The Art of Ecology: Writings of G. Evelyn Hutchinson*.
New Haven, CT: Yale University Press.

Iles, A., and M. J. Mulvihill. 2012. "Collaboration across Disciplines for Sustainability: Green Chemistry as an Emerging Multistakeholder community." *Environmental Science and Technology* 46:5643–49.

Jameson, Fredric. 2000. "'If I Find One Good City I Will Spare the Man': Realism and Utopia in Kim Stanley Robinson's Mars Trilogy." In *Learning from Other Worlds: Estrangement, Cognition, and the Politics of Science Fiction and Utopia*, edited by Patrick Parrinder, 208–232. Liverpool, UK: Liverpool University Press.

Jensen, Casper Bruun. 2015. "Mekong Scales: Domains, Test-Sites, and the Microuncommons." Paper presented at the Sawyer seminar workshop "Uncommons," University of California, Davis, May 28–30.

Kallis, Giorgos. 2019. *Limits: Why Malthus Was Wrong and Why Environmentalists Should Care*. Stanford, CA: Stanford University Press.

Kingsland, Sharon E. 2005. *The Evolution of American Ecology, 1890–2000*. Baltimore: Johns Hopkins University Press.

Klein, Ursula. 2013. "Materiality and Abstraction in Modern Chemistry." In J.-P. Llored, ed., *The Philosophy of Chemistry: Practices, Methodologies, and Concepts*, 342–62. Newcastle upon Tyne, UK: Cambridge Scholars Publishing.

Krebs, Charles J. 2016. *Why Ecology Matters*. Chicago: University of Chicago Press.

Landecker, Hannah. 2016. "Antibiotic Resistance and the Biology of History." *Body and Society* 22 (4): 19–52. https://doi.org/10.1177/1357034X14561341.

Landecker, Hannah. 2011. "Food as Exposure: Nutritional Epigenetics and the New Metabolism." *BioSocieties* 6 (2): 167–94.

Leane, Elizabeth. 2002. "Chromodynamics: Science and Colonialism in Kim Stanley Robinson's Mars Trilogy." *Ariel* 33 (1): 83–104.

Lee, Rachel. 2020. "Chemical Entanglements: Gender and Exposure [Special Issue]." *Catalyst: Feminism, Theory, Technoscience* 6 (1).

Lee, Stan, and Steve Ditko. 1962. *Amazing Fantasy No. 15*. New York: Marvel Comics.

Lefèvre, Wolfgang, and Ursula Klein. 2007. *Materials in Eighteenth-Century Science: A Historical Ontology*. Cambridge, MA: MIT Press.

Liboiron, M., M. Tironi, and N. Calvillo. 2018. "Toxic Politics: Acting in a Permanently Polluted World." *Social Studies of Science* 48 (3): 331–49. https://doi .org/10.1177/0306312718783087.

Lilley, Simon, and Dimitris Papadopoulos. 2014. "Material Returns: Cultures of Valuation, Biofinancialisation and the Autonomy of Politics." *Sociology* 48 (5): 972–88. https://doi.org/10.1177/0038038514539206.

Linebaugh, Peter. 2008. *The Magna Carta Manifesto: Liberties and Commons for All*. Berkeley: University of California Press.

Linebaugh, Peter. 2010. "Some Principles of the Commons." *Counterpunch*, January 8. https://www.counterpunch.org/2010/01/08/some-principles-of-the -commons/.

Linthorst, J. A. 2010. "An Overview: Origins and Development of Green Chemistry." *Foundations of Chemistry* 12 (1): 55–68.

Llored, Jean-Pierre, ed. 2013. *The Philosophy of Chemistry: Practices, Methodologies, and Concepts*. Newcastle upon Tyne, UK: Cambridge Scholars Publishing.

Logar, Nathaniel. 2011. "Chemistry, Green Chemistry, and the Instrumental Valuation of Sustainability." *Minerva* 49 (1): 113–36.

Lokesh, Kadambari, Luana Ladu, and Louise Summerton. 2018. "Bridging the Gaps for a 'Circular' Bioeconomy: Selection Criteria, Bio-Based Value Chain and Stakeholder Mapping." *Sustainability* 10 (6): 1–24. https://doi.org/10.3390/su10061695.

Lotringer, Sylvère, and Christian Marazzi, eds. 1980. *Autonomia: Post-political Politics*. New York: Semiotext(e).

Macauley, David. 2010. *Elemental Philosophy: Earth, Air, Fire, and Water as Environmental Ideas*. Albany: SUNY Press.

Maguire, Muireann. 2013. "In The Zone: The Strugatskii Brothers and the Poetics of Pollution in Russian Science Fiction." In *Literature and Chemistry: Elective Affinities*, edited by Margareth Hagen and Margery Vibe Skagen, 151–63. Aarhus, Denmark: Aarhus University Press.

Matus, K. 2010. "Policy Incentives for a Cleaner Supply Chain: The Case of Green Chemistry." *Journal of International Affairs* 64 (1): 121–36.

Mazomenos, Basilios E., and G. E. Haniotakis. 1981. "A Multicomponent Female Sex Pheromone of *Dacus oleae* Gmelin: Isolation and Bioassay." *Journal of Chemical Ecology* 7 (2): 437–44.

Mazomenos, Basilios E., and G. E. Haniotakis. 1985. "Male Olive Fruit Fly Attraction to Synthetic Sex Pheromone Components in Laboratory and Field Tests." *Journal of Chemical Ecology* 11 (3): 397–405.

Mazomenos, Basilios E., Anastasia Pantazi-Mazomenou, and Dimitra Stefanou. 2002. "Attract and Kill of the Olive Fruit Fly Bactrocera oleae in Greece as a Part of an Integrated Control System." *International Organization for Biological Control (IOBC-WPRS) Bulletin* 25:1–11.

McGurty, Eileen Maura. 2000. "Warren County, NC, and the Emergence of the Environmental Justice Movement: Unlikely Coalitions and Shared Meanings in Local Collective Action." *Society and Natural Resources* 13 (4): 373–87. https://doi.org/10.1080/089419200279027.

McManus, M. C., and C. M. Taylor. 2015. "The Changing Nature of Life Cycle Assessment." *Biomass and Bioenergy* 82:13–26. https://doi.org/10.1016/j.biombioe.2015.04.024.

Murphy, Michelle. 2006. *Sick Building Syndrome and the Problem of Uncertainty: Environmental Politics, Technoscience, and Women Workers*. Durham, NC: Duke University Press.

Murphy, Michelle. 2017. "Alterlife and Decolonial Chemical Relations." *Cultural Anthropology* 32 (4): 494–503. https://doi.org/10.14506/ca32.4.02.

Odum, Eugene P. 1953. *Fundamentals of Ecology*. Philadelphia: Saunders.

Papadopoulos, Dimitris. 2011. "The Imaginary of Plasticity: Neural Embodiment, Epigenetics and Ecomorphs." *Sociological Review* 59 (3): 432–56. https://doi.org/10.1111/j.1467-954X.2011.02025.x.

Papadopoulos, Dimitris. 2018. *Experimental Practice: Technoscience, Alterontologies, and More Than-Social Movements*. Durham, NC: Duke University Press.

Papadopoulos, Dimitris, Niamh Stephenson, and Vassilis Tsianos. 2008. *Escape Routes: Control and Subversion in the 21st Century*. London: Pluto.

Paracelsus. 1996. *Four Treatises of Theophrastus von Hohenheim, called Paracelsus*. Edited by Henry E. Sigerist. Translated by C. Lilian Temkin, George Rosen, Gregory Zilboorg, and Henry E. Sigerist. Baltimore: Johns Hopkins University Press.

Peijnenburg, Willie J. G. M. 2009. "Ecological Chemistry." In *Environmental and Ecological Chemistry*, vol. 3, edited by Aleksandar Sabljic, 45-84. Oxford: Eolss.

Peters, John Durham. 2015. *The Marvelous Clouds: Toward a Philosophy of Elemental Media*. Chicago: University of Chicago Press.

Pignarre, Philippe, and Isabelle Stengers. 2011. *Capitalist Sorcery: Breaking the Spell*. Basingstoke, UK: Palgrave Macmillan.

Poliakoff, Martyn. 2014. "Is It Time for an F-factor?" *Green Chemistry* 16:21–22.

Poliakoff, Martyn, P. Licence, and M. W. George. 2018. "A New Approach to Sustainability: A Moore's Law for Chemistry." *Angewandte Chemie International Edition* 57 (39): 12590–91. https://doi.org/10.1002/anie.201804004.

Pratt, Mary Louise. 1992. *Imperial Eyes: Travel Writing and Transculturation*. London: Routledge.

Puig de la Bellacasa, María. 2010. "Ethical Doings in Naturecultures." *Ethics, Place and Environment: A Journal of Philosophy and Geography* 13 (2): 151–69.

Puig de la Bellacasa, María. 2015. "Making Time for Soil: Technoscientific Futurity and the Pace of Care." *Social Studies of Science* 45 (5): 691–716. https://doi .org/10.1177/0306312715599851.

Puig de la Bellacasa, María. 2017. *Matters of Care: Speculative Ethics in More than Human Worlds*. Minneapolis: University of Minnesota Press.

Reid, Herbert G., and Betsy Taylor. 2010. *Recovering the Commons: Democracy, Place, and Global Justice*. Champaign: University of Illinois Press.

Roberts, Jody A. 2010. "Reflections of an Unrepentant Plastiphobe: Plasticity and the STS Life." *Science as Culture* 19 (1): 101–20. https://doi.org/10.1080 /09505430903557916.

Roberts, Jody Alan. 2005. "Creating Green Chemistry: Discursive Strategies of a Scientific Movement." PhD diss., Virginia Polytechnic Institute and State University. https://vtechworks.lib.vt.edu/handle/10919/27529.

Robinson, Kim Stanley. 1992. *Red Mars*. London: HarperCollins.

Robinson, Kim Stanley. 1993. *Green Mars*. London: HarperCollins.

Robinson, Kim Stanley. 1996. *Blue Mars*. London: HarperCollins.

Rockström, Johan, Will Steffen, Kevin Noone, Åsa Persson, F. Stuart Chapin III, Eric Lambin, Timothy M. Lenton, Marten Scheffer, Carl Folke, Hans Joachim Schellnhuber, Björn Nykvist, Cynthia A. de Wit, Terry Hughes, Sander van der Leeuw, Henning Rodhe, Sverker Sörlin, Peter K. Snyder, Robert Costanza, Uno Svedin, Malin Falkenmark, Louise Karlberg, Robert W. Corell, Victoria J. Fabry, James Hansen, Brian Walker, Diana Liverman, Katherine Richardson,

Paul Crutzen, and Jonathan Foley. 2009. "Planetary Boundaries: Exploring the Safe Operating Space for Humanity." *Ecology and Society* 14 (2): 32–65.

Sanderson, K. 2011. "Chemistry: It's Not Easy Being Green." *Nature* 469 (7328): 18–20. https://doi.org/10.1038/469018a.

Shadaan, Reena, and Michelle Murphy. 2020. "Endocrine-Disrupting Chemicals (EDCs) as Industrial and Settler Colonial Structures: Towards a Decolonial Feminist Approach." *Catalyst: Feminism, Theory, Technoscience* 6 (1): 1–36.

Shapiro, Nicholas. 2015. "Attuning to the Chemosphere: Domestic Formaldehyde, Bodily Reasoning, and the Chemical Sublime." *Cultural Anthropology* 30 (3): 368–93. https://doi.org/10.14506/ca30.3.02.

Sheldon, Roger A. 1992. "Organic Synthesis: Past, Present and Future (Advantages of Incorporating Catalysis to Organic Synthesis)." *Chemistry and Industry* (December 7): 903–6.

Sheldon, Roger A. 2007. "The E Factor: Fifteen Years On." *Green Chemistry* 9 (12): 1273–83. https://doi.org/10.1039/b713736m.

Sheoin, Tomás Mac, and Frank Pearce, eds. 2014. "Bhopal and After." Special issue, *Social Justice* 41 (1–2).

Stableford, Brian. 2005. "Science Fiction and Ecology." In *A Companion to Science Fiction*, edited by David Seed, 127–41. Malden, MA: Blackwell.

Stengers, Isabelle. 2005. "An Ecology of Practices." *Cultural Studies Review* 11 (1): 183–96.

Strathern, Marilyn. 2000. "Environments Within: An Ethnographic Commentary on Scale." In *Culture, Landscape, and the Environment: The Linacre Lectures 1997*, edited by Kate Flint and Howard Morphy, 44–71. Oxford: Oxford University Press.

Strugatsky, Arkady, and Boris Strugatsky. 1977. *Roadside Picnic/Tale of the Troika*. New York: Macmillan.

Tarkovsky, Andrey, dir. 1979. *Stalker*. Soviet Union: Mosfilm.

Tarkovsky, Andrey. 1989. *Sculpting in Time*. Translated by Kitty Hunter-Blair. Austin: University of Texas Press.

Tiptree, James, Jr. 1975. "Love Is the Plan the Plan Is Death." In *Warm Worlds and Otherwise*. New York: Ballantine.

Trost, Barry. 1991. "The Atom Economy: A Search for Synthetic Efficiency." *Science* 254:1471–77. https://doi.org/10.1126/science.1962206.

Tsing, Anna. 2012. "On Nonscalability: The Living World Is Not Amenable to Precision-Nested Scales." *Common Knowledge* 18 (3): 505–24. https://doi.org/10.1215/0961754x-1630424.

Tsing, Anna. 2015. *The Mushroom at the End of the World: On the Possibility of Life in Capitalist Ruins*. Princeton, NJ: Princeton University Press.

Wall, Derek. 2014. *The Commons in History: Culture, Conflict, and Ecology*. Cambridge, MA: MIT Press.

Whitehead, Alfred North. 1925. *Science and the Modern World*. New York: Macmillan.

Whitehead, Alfred North. (1929) 1979. *Process and Reality: An Essay in Cosmology.* Edited by D. R. Griffin and D. W. Sherburne. New York: Free Press.

Wilson, Rob. 2015. "Toward an Ecopoetics of Oceania: Worlding the Asia-Pacific Region as Space-Time Ecumene." In *American Studies as Transnational Practice: Turning toward the Transpacific*, edited by Yuan Shu and Donald E. Pease, 213–36. Hanover, NH: Dartmouth College Press.

Woodhouse, E. J., and Steve Breyman. 2005. "Green Chemistry as Social Movement?" *Science, Technology and Human Values* 30 (2): 199–222. https://doi.org/10.1177/0162243904271726.

Wylie, Sara. 2012. "Hormone Mimics and Their Promise of Significant Otherness." *Science as Culture* 21 (1): 49–76. https://doi.org/10.1080/09505431.2011.566920.

Wylie, Sara. 2017. *Fractivism.* Durham, NC: Duke University Press.

3

ELEMENTARY FORMS OF ELEMENTARY FORMS: OLD, NEW, AND WAVY

Stefan Helmreich

Dmitri Mendeleev sketched the first draft of his periodic table of the elements (figure 3.1) on the back of a letter he received in 1869 from the owner of a cooperative cheese-making factory, a factory that Mendeleev, as a big fan of agricultural modernization, had recently been invited to tour. After finishing a full version of the table—still on view, alongside his rough preliminary attempt, in the Mendeleev Memorial Museum Apartment in Saint Petersburg—Mendeleev took the tour and wrote a report on how the cooperative farm could help recently emancipated Russian serfs organize the elements of production (cattle, feed, milk) toward making reliably regular, standardized, salable cheese.[1] In other words, Mendeleev—whether in the lab, in the study, or on the farm—was thinking about elements and how they might be arranged and combined.

Take Mendeleev as one orienting figure for an early history of *elements thinking*, that pivot in contemporary social theory toward understanding amalgams of natural and cultural objects through their chemical connections and relations, their molecular-molar meshwork,

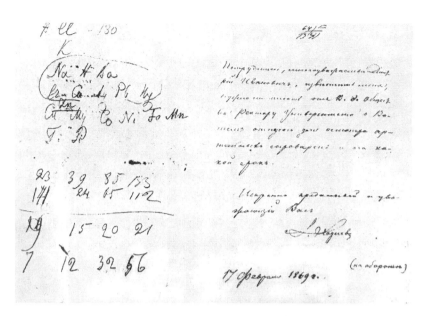

FIGURE 3.1: Mendeleev's first sketch for the periodic table of the elements, on view at the Mendeleev Memorial Museum Apartment in Saint Petersburg. Photograph by Stefan Helmreich.

and their material and mediated substantiations. Contemporary writers in environmental history, ecocriticism, and anthropology have recently experimented with (re)activating pre-Socratic philosopher Empedocles's fifth-century BCE four-element cosmogony (which had the world made of fire, air, water, and earth) or sometimes the classical Chinese *wuxing* system of five agents/phases/elements (wood, fire, earth, metal, and water) to offer new angles on ecological process and politics.[2] Other scholars, hewing to a more Mendeleevian line, have started to follow the social life of such elements as carbon, helium, plutonium, lead, and nitrogen as well as such compounds as formaldehyde and methamphetamine.[3] In so doing, these authors have sought to be technoscientifically precise in their response to recent calls in social theory to attend to the materiality and agency of the nonhuman and more-than-human world. One difference between this latter elements thinking and Mendeleev's, of course, is that in today's work, chemical and social bonds are no longer figured simply as loosely analogous to one another but as constantly imbricated in and transformative of one another.

Still, Mendeleev provides a useful starting point for a genealogy for elements thinking, one that reaches further back than the past ten or so years of the elemental turn. Think, too, of Dmitri Mendeleev's 1869 publication of the periodic table of the elements alongside sociologist Émile Durkheim's 1912 *Elementary Forms of the Religious Life*. During the later nineteenth and early twentieth centuries, writers across natural and social fields of inquiry were outlining such things as the first principles of thermodynamics (Rankine 1859), the origin of species (Darwin 1859), the emergence of primitive culture (Tylor 1871), and the organization of ancient law and society (Maine 1861; Morgan 1877). The aspiration across all these arenas of investigation was to scale up—one might say, to *compound up*—from elementary to more complex processes and forms.

There were some precedents too. Take, for example, French chemist Antoine Lavoisier, who, in his 1789 *Traité élémentaire de chimie* (Elementary treatise of chemistry), posited that what made an element an element was its status as a substance that could not be further broken down by any technique of chemical analysis—or, to be more exact, by any technique *known at the time of the inquiry*. Lavoisier's framework was, famously, a provisional one, ready to admit future discoveries, and it was inspired by the French epistemologist Etienne Bonnot de Condillac's 1778 textbook, *Logic*, which suggested that human language, properly structured and disciplined, could disclose the order of nature by properly guiding analysis and by formatting in-the-world (or in-the-lab) instrumental and experimental practice (Roberts 1992). Lavoisier's model, which posed elements both as things in the world *and* as results of a rational scheme of predictive representation (and intervention [see Hacking 1983]), would set the stage for the approach taken by Mendeleev in his framing of the periodic table. In philosophy, Denis Diderot's (1774–80) *Elements of Physiology* extended such elements-and-compounds thinking to living matter, offering particularly to the human sciences a vision of human bodies as animated by a reductionist, yet compounding, materialist vitalism.

Return, however, to Émile Durkheim. What were "elements" for Durkheim? In *The Rules of Sociological Method*, Durkheim ([1895] 1982) described social facts as sui generis, not derivable from, say, psychological facts (departing here from the vitalist reductionism of Diderot). Social facts "lie outside the consciousness of individuals as such in the same way as the distinctive features of life lie outside the chemical substances that make up a living organism" (40). Durkheim called upon the properties of water to elaborate his analogy: "The liquidity of water, its sustaining and other properties, are not in the two gases of which it is composed, but in the complex substance which they form by coming together" (39). In other words, social facts, like liquids, may depend upon their constituent elements, but their properties are not derivable from such elements.

The very notion of an element, however, was a mobile, scale-shifting one for Durkheim. In *Rules*, he also wrote, "We know that societies are made up of a number of parts added on to each other. Since the nature of any composite necessarily depends upon the nature and number of the elements that go to make it up and the way in which these are combined, these characteristics are plainly what we must take as our basis" (111). These elements, however, were *not* individual psyches, but other social forms. Durkheim explained, "It is known in fact that the constituent parts of every society are themselves societies of a simpler kind" (112). In other words, the "social" became its own element, irreducible. Durkheim *did* at times posit a simplest social element, which he called "the horde" (113), or sometimes "primitive horde" (compare Hayden on crowds, this volume). But hordes added to hordes just made new social things—in this case, segmentary clan systems. Likewise, for the Durkheim of 1893's *The Division of Labor*, while combinations of elements led to qualitatively different kinds of solidarity—mechanical and organic—these were still *social*.

Political theorist Paul Q. Hirst argues that such "elements" in Durkheim operate differently from "elements" in periodic table thinking: "Durkheim's notion of the element or atom is like the speculative construction of the constituents of the universe in Naturphilosophie (Democritus or eighteenth-century mechanical materialism). The theoretical conditions of the existence of 'elements' in chemistry are quite different. The order and the relation of elements is determined by a theoretical construction, the Mendeleev table of the elements. This theoretical construction is not the recording of the results of 'discovery' in an orderly form, it is a means of *producing* these phenomena of chemistry, of realizing the objects which it unites" (Hirst [1975] 2010, 126n8; emphasis added). In his 1940 book *The Philosophy of No*, Gaston Bachelard made a similar point: "In the Mendeleev organization of elementary substances, it becomes apparent that, little by little, *law surpasses fact*, that the *order* of substances imposes itself as rationality" (Hirst [1975] 2010, 127n8). In other words, new chemical elements become thinkable through the theory; they are not so much discovered as *envisaged*. That envisaging, coupled with experiment, can even be mobilized to create—and not just predict—new elements: representing, intervening, making.

And here it seems to me that Durkheim's elements *do* have something in common with those on the periodic table. Durkheim's conceptual apparatus is *also* "determined by a theoretical construction"—namely, one that posits social organization as a composing logic. And, as the twentieth century has shown us, the laws of social sciences *can* often surpass fact, imposing themselves as structures of rationality and control and as the order of the factual world—as Max Weber worried when he proposed the oppressive figure of modernity's "iron cage" in his *Protestant Ethic and the Spirit of Capitalism* (see Baehr 2001).

SO: FAST FORWARD TO TODAY, and elements thinking is either similar (playing on the capaciousness and underdetermination of the concept of the element, classical, chemical, combinatoric) or is somewhat different (calling upon elements quite precisely as chemopolitical commodities [e.g., Mukharji 2016], as animating "dangerscapes" of particulate toxics [see, e.g., Mitman, Murphy, and Sellers 2004], and as molecular mash-ups of human practice and posthuman being from the bodily to the global). But what might most distinguish early twenty-first-century elements thinking from that of the late nineteenth century is a refusal of divisions between science and culture, inorganic and organic, and an attention to hybrids, chimeras, and material-symbolic mixtures—as well as their inextricable multispecies politics.

In a now-classic reading of the periodic table as a chart of the "kinship relations of the elements," Donna Haraway in 1997 wrote that such relations were "a natural-technical object of knowledge that semiotically and instrumentally puts terrans in their proper place" (54). More, she offered, the "transuranic elements"—those made, not born—"have forced humans to recognize their problematic kinship with each other as fragile earthlings at a scale of shared vulnerability and mortality" (55). Where eighteenth- and nineteenth-century elements thinking was about constituent parts of coherent systems, twentieth- and twenty-first-century elements thinking is about partial and fraught relational connections in incomplete assemblages, as well as about the qualities (e.g., toxicities) that are called forth by those assemblages. If "elementary structures of kinship" for Claude Lévi-Strauss ([1949] 1969) referred to such forms as cross-cousin marriage, which he argued was logically forced by some kin systems (and which Gayle Rubin [1975] argued was actually the result of naturalizing heterosexual reproduction and marriage), the new elemental kinship might offer different affinities of what Haraway has called "friendship, work, partially shared purpose, intractable collective pain, and persistent hope" (1997, 265). The shared substances in the mix here include such elements and compounds as "carbon, rare earth minerals, and polymetallic sulfides" (Parikka 2016), moving across many organic kinds of bodies (see Choy on "breathing with," this volume; Dumit on bromine as self and other, this volume; Murphy on chemical and political affinities and anti-affinities, this volume) and, indeed, through the reticulated pathways of local ecologies as well as of the whole planet (Alaimo 2016; Papadopoulos, this volume; Puig de la Bellacasa, this volume; Weston 2017)—fire, air, water, earth.

LET ME CALL UPON and torque both old and new genres of elements thinking, then, by thinking with water, both as an irreducible element (à la Empedocles's *water* and Durkheim's *social*) and as a compound substance that can carry chemopolitical elements into combination (see Chen, MacLeod, and Neimanis 2013). Consider Eva Hayward's (2014) work on *transxenoestrogenesis*, which tracks how manufactured estrogens travel into aquatic worlds and morph the hormonal bodies of frogs, turtles, and fish, bringing old and new elementalities together. Consider Melody Jue's call for thinking about *ocean media*, in connection with which she writes, "The ocean is not a generic fluid but an environment of particular chemical composition" (2014, 89). Consider Astrida Neimanis, who, in *Bodies of Water: Posthuman Feminist Phenomenology* (2017), elaborates and specifies what such ocean media thinking might mean for feminist elemental thinking, writing of how the "flow of biomagnified toxins in breast milk" (33) might be understood as a key moment within a damaged hydrosocial cycle. Or consider Joseph Masco's report (this volume) on the circulation of plutonium in the seas (and see Neimanis, Neimanis, and Åsberg 2017 on chemical weapons "buried" on the seafloor—chemical elementalities that exist as "latent" toxins). How do such chemoelementalities travel into and through the old-school elementality of the ocean? The question of "travel" or motion is important because it is *in* movement that fresh connections and recombinations are made and unmade. How can heterogeneous genres of elements be tracked through the elementality of water?

One answer is through the forms known as *waves*. If fluidity and flow are deformations that give form to classical understandings of water as an element, waves make up one texture of that element. David Macauley's *Elemental Philosophy: Earth, Air, Fire, and Water as Environmental Ideas* (2010) suggests that one of the "archetypical patterns" taken by water is the wave, a "hallmark" of its "restless movement" (47). John Durham Peters's *The Marvelous Clouds: Towards a Philosophy of Elemental Media* (2015) takes waves (of water and sound) as phenomena that facilitate and disturb ocean-borne communication, of ships and cetaceans. Jeffrey Jerome Cohen and Lowell Duckert's *Elemental Ecocriticism: Thinking with Earth, Air, Water, and Fire* (2015), while starting with Empedoclean elements, does so in the spirit of a "productive anachronism." Cohen and Duckert argue that a straightforward appropriation of old-style elements cannot quite get at ecocritical concerns with extinction, climate change, toxins, and slow and fast environmental violence. In their 2017 edited volume, *Veer Ecology*, Cohen and Duckert point out that the word *environ*, which derives from the French *virer*, "to turn"—pointing therefore to various animal, nonhuman, chemical turns—puts the "swirl" and "turbulent whirl" of the world into elemental ecocriticism.

So, let me present a slowly swirling and whirling wave under new investigation by oceanographers these days: the internal wave. While I could have chosen another sort of wave—perhaps the breaking open ocean wave, which mixes together air and water—the internal wave opens into a story I want to use to exemplify the possibilities of old and new elemental thinking.

I learned about internal waves as I was doing ethnographic fieldwork among oceanographers at MIT and at the Scripps Institution of Oceanography (for oceanographic papers on internal waves, see Alford, Klymak, and Carter 2014; Peacock 2014). Internal waves oscillate beneath the surface of the sea, in ocean water stratified by salinity or temperature. Such submarine waves can be hundreds of feet tall and can even break. Even if they are linear waves—forms moving through water without transporting water itself—they can gradually transport animal larvae, plankton, and chemical mixtures, as waves compress and rarefy seawater and nudge substances along (see Lennert-Cody and Franks 2002). A biological oceanographer at Scripps explained to me how the troughs and crests of internal waves might form a kind of mobile ecological context for such biochemical relays. He went on to tell me that understanding such relays often required an understanding of ratios between such properties as viscosity (a fluid's resistance to deformation) and molecular diffusion—in other words, between the bulk elementality of water and the elementality of the molecules that constitute and combine with it (on viscosity, see Couling and Hein 2019).

What chemical relays are carried and combined on internal waves? In California, just off the coast of Palos Verdes Peninsula, an undersea location that the *Los Angeles Times* once designated as "the nation's largest ocean dumping ground of DDT" (Ferrell 1992), internal waves have been found to stir up sedimented DDT as well as PCB contamination, the material legacy of DDT manufacture and dumping from 1947 to 1983 by the Montrose Chemical Corporation. Both DDT and PCBs (polychlorinated biphenyls) have worked their way through microbes, fish, seabirds, and marine mammals—and have done so in ways facilitated by the generation of internal waves off the Palos Verdes Shelf, which have suspended such toxic elementalities and molecularities higher up in the water than they might otherwise be (see Ferré, Sherwood, and Wiberg 2010).

How do oceanographers come to generate such knowledge claims? The answer is through in-the-field measurement and monitoring, to be sure—but also in the laboratory, through modeling miniature internal waves. At MIT, I followed one such experiment in lab replication of internal wave phenomena. During the experiment I witnessed, a doctoral student sought to create a model of an internal wave in a tank of water that was five meters long, one meter deep, and a half-meter wide. He filled the tank with a mix of fresh and salt water—a mix of elementalities—with the aim of creating a depth of water that might have a gradient of densities analogous to what one might expect to see in the ocean (see figure 3.2). He would then "run" a wave through the medium. One of the puzzles with which scientists like this graduate student must grapple is how to make a scale model, how to scale down ocean water. As he put it to me, "What we've done here is scale down the ocean. If you were a little man swimming down into the tank, you'd notice it would get saltier and saltier." The gradient of salt water is important for how the internal wave will travel—and, if one were interested in the larger sorts of waves that this "lab-wave" was meant to model, for how substances could be transported through the ocean, understood as an undulating ecology. If one were "a little man," the water would also feel more viscous—perhaps giving one the feeling of "sailing on the seas of cheese," as progressive rock band Primus put it on their 1991 album of that title, calling atten-

FIGURE 3.2: A wave tank, being filled with salt and fresh water in advance of an experimental wave being run through the water medium. Photograph by Stefan Helmreich.

tion to the sometime comedy of humans thinking across impossible-to-experience scales.[4] Changing scales is often, also, about changing the *qualities* of the substances—the elements—in play.

Once upon a time, in 1879, well before lab experiments such as the one I have just recounted unfolded, the applied mathematician Sir Horace Lamb argued in his foundational text, *Hydrodynamics*—a sort of elementary forms of waves textbook—that "we assume that the properties of the smallest portions into which we can conceive [the matter of fluid] to be divided are the same as those of the substance in bulk" (1879 [1895], 1). Nowadays, oceanographers know that making a scale model is *not* so simple; making an internal miniwave requires controlling for such properties as viscosity—which does not scale down easily (see Latour 1987, 239) and is, as Nancy Couling and Carola Hein (2019,) note, "inherently relational."

Lamb's claim about the similarity of water at all scales is similar to those that Durkheim made about the "social," which offered the notion that the *smallest portion of the social* might behave similarly to the *largest kind of the social* (which was, after all, the basic argument in *The Elementary Forms of the Religious Life*, which held that all societies produced religions of the same form, as a social arrangement organized around prohibitions around the sacred). When Durkheim, in *Suicide* (1897), mapped suicide rates as "waves" through the elementalities of social life, sociality as such was not undone as the underlying analytic—as liquidity, contra Lamb, has been by today's water and wave modeling, and as elements thinking in our post-Harawavian days has done to anything as sure as the social. Old and new elementalities, considered together, have produced new compounds, new phenomena, analytics, and theories essential to mix into our accounts of the relations and bonds that twine together the forms and materials of our multiplicitous worlds.

ACKNOWLEDGMENTS

My thanks to Xenia Cherkaev for tracking down the specifics of the Mendeleev story; to Heather Paxson for helping me see this episode as one in the long history of industrial cheese production; to Nikolai Ssorin-Chaikov for accompanying Xenia, Heather, and me to the Mendeleev Memorial Museum Apartment in Saint Petersburg; and to our historian docent, Tatiana Viktorovna Strekopytova, for essential guidance into Mendeleev's world. Thanks also to the editors of the present book—Natasha Myers, Dimitris Papadopoulos, and María Puig de la Bellacasa—for clarifying prods. I also thank my fellow panelists on the 2016 panel at the meetings of the Society for the Social Studies of Science in Barcelona, from which this collection grew: Tim Choy, Joe Dumit, Joe Masco, Michelle Murphy, and Astrid Schrader. Cori Hayden, unable to make it to the panel itself, was a vital interlocutor during the further writing and revision. I also thank Will Deringer for putting me on the trail of the Lavoisier and Diderot stories. Thanks also to participants at Elements: Matters, Analytics, and Worlds, a collaborative research residency held in 2019 at the University of California Humanities Research Institute on the campus of the University of California, Irvine; organizers Mei Zhan and Daniel Fisher, along with participants Nikhil Anand, Andrea Ballestero, Tim Choy, and John Durham Peters, provided crucial input.

1. See Mendeleev 1869.

2. On thinking about contemporary environmental questions through the classical Greek elements, see, to begin, Cohen and Duckert 2015; Macauley 2010; Peters 2015. Four-element thinking has arrived in anthropology through work on fire (e.g., Fisher, 2020; Neale 2018), air (e.g., Choy 2012), water (too many to mention—the conversation is a long-standing one—but useful touchstones are Ballestero 2019; Chen, MacLeod, and Neimanis 2013; Strang 2005), and earth (again, there are many data points, but see Ballestero 2016, 2019). On the Chinese *wuxing* system, see Zhan, forthcoming.

3. On carbon, see Günel 2019; Whitington 2016; on helium, see McCormack 2018; on plutonium, see Masco 2004, this volume; on lead, see Fennell 2016, Montoya 2017; on petroleum, see Povinelli 2016; on nitrogen, see Morris, forthcoming; on formaldehyde, see Shapiro 2015; on methamphetamine, see Pine 2019.

4. The strange comedy of thinking in terms of four elements today, in our secular age, finds an example in the work of children's author Rick Riordan (2010), who has a character say, "Ah, Mastery of the Five Elements! . . . How to tame the five essential elements of the universe—earth, air, water, fire, and cheese!" (231).

REFERENCES

Alford, Matthew H., Jody M. Klymak, and Glenn S. Carter. 2014. "Breaking Internal Lee Waves at Kaena Ridge, Hawaii." *Geophysical Research Letters* 41:906–12.

Alaimo, Stacy. 2016. *Exposed: Environmental Politics and Pleasures in Posthuman Times*. Minneapolis: University of Minnesota Press.

Bachelard, Gaston. (1940) 1968. *The Philosophy of No: A Philosophy of the New Scientific Mind*. Translated by G. C. Waterson. New York: Orion.

Baehr, Peter. 2001. "The 'Iron Cage' and the 'Shell as Hard as Steel': Parsons, Weber, and the *Stahlhartes Gehäuse* Metaphor in *The Protestant Ethic and the Spirit of Capitalism*." *History and Theory* 40 (2): 153–69.

Ballestero, Andrea. 2016. "Spongy Aquifers, Messy Publics." *Limn* 7. https://limn .it/articles/spongy-aquifers-messy-publics/.

Ballestero, Andrea. 2019. *A Future History of Water*. Durham, NC: Duke University Press.

Chen, Cecilia, Janine MacLeod, and Astrida Neimanis, eds. 2013. *Thinking with Water*. Montreal: McGill-Queen's University Press.

Choy, Tim. 2012. "Air's Substantiations." In *Lively Capital*, edited by Kaushik Sunder Rajan, 121–52. Durham, NC: Duke University Press.

Cohen, Jeffrey Jerome, and Lowell Duckert, eds. 2015. *Elemental Ecocriticism: Thinking with Earth, Air, Water, and Fire*. Minneapolis: University of Minnesota Press.

Cohen, Jeffrey Jerome, and Lowell Duckert, eds. 2017. *Veer Ecology: A Companion for Environmental Thinking*. Minneapolis: University of Minnesota Press.

Condillac, Etienne Bonnot, Abbé de. (1778) 1780. *La logique, ou Les premiers développements de l'art de penser: Ouvrage* élémentaire. Paris: Edité par L'Esprit et de Bure l'ainé.

Couling, Nancy, and Carola Hein. 2019. "Viscosity." *Society and Space: Volumetric Sovereignty* Part 3: *Turbulence*, https://www.societyandspace.org/articles /viscosity.

Darwin, Charles. (1859) 1964. *On the Origin of Species*. Cambridge, MA: Harvard University Press.

Diderot, Denis. (1774–80) 1964. *Eléments de physiologie: édition critique, avec une introduction et des notes par J. Mayer*. Paris: Didier.

Durkheim, Émile. (1893) 1984. *The Division of Labor in Society*. Translated by W. D. Halls. New York: Free Press.

Durkheim, Émile. (1895) 1982. *The Rules of Sociological Method, and Selected Texts on Sociology and Its Method*. Edited by Steven Lukes. Translated by W. D. Halls. New York: Free Press.

Durkheim, Émile. (1897) 1952. *Suicide: A Study in Sociology*. Translated by John A. Spaulding and George Simpson. London: Routledge.

Durkheim, Émile. (1912) 1915. *The Elementary Forms of the Religious Life*. Translated by Joseph Ward Swain. London: Allen and Unwin.

Fennell, Catherine. 2016. "Are We All Flint?" *Limn* 7. https://limn.it/articles/are -we-all-flint/.

Ferré, Bénédicte, Christopher R. Sherwood, and Patricia L. Wiberg. 2010. "Sediment Transport on the Palos Verdes Shelf, California." *Continental Shelf Research* 30:761–80.

Ferrell, David. 1992. "Off Palos Verdes, a DDT Dumping Ground Lingers." *Los Angeles Times*, September 9.

Fisher, Daniel. 2020. "Fire." In *Anthropocene Unseen: A Lexicon*, edited by Anand Pandian and Cymene Howe, 191–95. Goleta, CA: Punctum.

Günel, Gökçe. 2019. *Spaceship in the Desert: Energy, Climate Change, and Urban Design in Abu Dhabi*. Durham, NC: Duke University Press.

Hacking, Ian. 1983. *Representing and Intervening: Introductory Topics in the Philosophy of Natural Science*. Cambridge, UK: Cambridge University Press.

Haraway, Donna, 1997. *Modest_Witness@Second_Millennium.FemaleMan©_Meets _OncoMouse™: Feminism and Technoscience*. New York: Routledge.

Hayward, Eva. 2014. "Transxenoestrogenesis." *Transgender Studies Quarterly* 1 (1–2): 255–58.

Hirst, P. Q. (1975) 2010. *Durkheim, Bernard, and Epistemology*. London: Routledge.

Jue, Melody. 2014. "Vampire Squid Media." *Grey Room* 57:82–105.

Lamb, Sir Horace. (1879) 1895. *Hydrodynamics*. Cambridge: Cambridge University Press.

Latour, Bruno. 1987. *Science in Action: How to Follow Scientists and Engineers through Society*. Cambridge, MA: Harvard University Press.

Lavoisier, Antoine. 1789. *Traité élémentaire de chimie* [Elementary treatise of chemistry]. Paris: Cuchet.

Lennert-Cody, E. Cleridy, and Peter J. S. Franks. 2002. "Fluorescence Patches in High-Frequency Internal Waves." *Marine Ecology Progress Series* 235:29–42.

Lévi-Strauss, Claude. (1949) 1969. *The Elementary Structures of Kinship.* Translated by J. Bell and J. von Sturmer. Boston: Beacon.

Macauley, David. 2010. *Elemental Philosophy: Earth, Air, Fire, and Water as Environmental Ideas.* Albany: SUNY Press.

Maine, Sir Henry Hames Sumner. 1861. *Ancient Law: Its Connection with the Early History of Society, and Its Relation to Modern Ideas.* London: John Murray.

Masco, Joseph. 2004. "Mutant Ecologies: Radioactive Life in Post–Cold War New Mexico." *Cultural Anthropology* 19 (4): 517–50.

McCormack, Derek P. 2018. *Atmospheric Things: On the Allure of Elemental Envelopment.* Durham, NC: Duke University Press.

Mendeleev, Dmitri. 1869. "About Cheese-Making Cooperatives." *Almanac of the Free Economic Society* 2 (6): 538–59.

Mitman, Gregg, Michelle Murphy, and Christopher Sellers, eds. 2004. *Landscapes of Exposure: Knowledge and Illness in Modern Environments.* Osiris 19. Chicago: University of Chicago Press.

Montoya, Teresa. 2017. "Yellow Water: Rupture and Return One Year after the Gold King Mine Spill." *Anthropology Now* 9 (3): 91–115.

Morgan, Lewis Henry. 1877. *Ancient Society, or Researches in the Lines of Human Progress from Savagery through Barbarism to Civilization.* London: Macmillan.

Morris, Christopher. Forthcoming. *Nitrogen Networks.*

Mukharji, Projit Bihari. 2016. "Parachemistries: Colonial Chemopolitics in a Zone of Contest." *History of Science* 54 (4): 362–82.

Neale, Timothy. 2018. "Digging for Fire: Finding Control on the Australian Continent." *Journal of Contemporary Archaeology* 5 (1): 79–90.

Neimanis, Astrida. 2017. *Bodies of Water: Posthuman Feminist Phenomenology.* London: Bloomsbury.

Neimanis, Astrida, Aleksija Neimanis, and Cecilia Åsberg. 2017. "Fathoming Chemical Weapons in the Gotland Deep." *Cultural Geographies* 24 (4): 631–38.

Parikka, Jussi. 2016. "Elements, a New Book Series Edited by Nicole Starosielski and Stacy Alaimo." *Machinology: Machines, Noise, and Some Media Archaeology* (blog). https://jussiparikka.net/2016/10/13/elements-a-new-book-series/.

Peacock, Thomas. 2014. "In Pursuit of Internal Waves." *Bulletin of the American Physical Society* 59 (20). http://meetings.aps.org/Meeting/DFD14/Session/K14.

Peters, John Durham. 2015. *The Marvelous Clouds: Towards a Philosophy of Elemental Media.* Chicago: University of Chicago Press.

Pine, Jason. 2019. *A Decomposition: Methlabs, Alchemy and the Matter of Life.* Minneapolis: University of Minnesota Press.

Povinelli, Elizabeth A. 2016. "Petroleum." Theorizing the Contemporary, *Fieldsights,* January 21. https://culanth.org/fieldsights/petroleum.

Rankine, William. 1859. "Principles of Thermodynamics." In *A Manual of the Steam Engine and Other Prime Movers*, 299–478. London: Charles Griffin.

Riordan, Rick. 2010. *The Red Pyramid*. New York: Hyperion.

Roberts, Lissa. 1992. "Condillac, Lavoisier, and the Instrumentalization of Science." *The Eighteenth Century* 33 (3): 252–71.

Rubin, Gayle. 1975. "The Traffic in Women: Notes on the 'Political Economy' of Sex." In *Toward an Anthropology of Women*, edited by Rayna Reiter, 157–210. New York: Monthly Review Press.

Shapiro, Nick. 2015. "Attuning to the Chemosphere: Domestic Formaldehyde, Bodily Reasoning, and the Chemical Sublime." *Cultural Anthropology* 30 (3): 368–93.

Strang, Veronica. 2005. "Common Senses: Water, Sensory Experience and the Generation of Meaning." *Journal of Material Culture* 10 (1): 92–120.

Tylor, Edward B. 1871. *Primitive Culture: Researches into the Development of Mythology, Philosophy, Religion, Language, Art, and Custom*. London: John Murray.

Weston, Kath. 2017. *Animate Planet: Making Visceral Sense of Living in a High-Tech Ecologically Damaged World*. Durham, NC: Duke University Press.

Whitington, Jerome. 2016. "Carbon." Theorizing the Contemporary, *Fieldsights*, April 6. https://culanth.org/fieldsights/carbon.

Zhan, Mei. Forthcoming. "Bringing Medicine Back to Life."

4

SUBSTANCE AS METHOD: BROMINE, FOR EXAMPLE

Joseph Dumit

> If we pay it due attention, we will find
> more in nature than what we observe at
> first glance.
> —Alfred North Whitehead,
> *The Concept of Nature*

In what he calls the instinctive attitude, Alfred North Whitehead ([1920] 2015) describes the essence of science and of everyday life. Isabelle Stengers (2011) explains that Whitehead means that the more we look at something, attend to it, and investigate it, the more we find in it. The substance of the thing changes before our eyes, surprisingly so. We notice things we had not even thought of before.

This instinctive attitude seems both completely right and somewhat puzzling. If it is so natural, why does it seem so rare? The opposite could also be true: we look at things every day, and they stay the same. Pretty soon we do not notice them at all. What is the difference then, and is there a way to cultivate the first? Are there processes that keep us enthralled to our existing concepts keeping us from seeing more?

In this chapter, I suggest that we might make a study of those who are surprised by a substance they have been carefully investigating: surprised to the extent that they have to invent new ways of talking about and working with the substance. They also cannot help but talk about the process, since they are equally struck by how long they had not been able to notice what is now obvious. And they recognize something new about themselves and the assumptions that they have been holding. They come to understand that the very concepts, metaphors, and methods they have been using were based on different substances. This is what can be called *substance as method*, an approach that suspects that each substance might be its own method and metaphor. Each substance, in other words, is elemental, and elements are composite (see Helmreich, this volume; Hayden, this volume). In some cases, this means that the substance resists being reduced to more elementary interactions or that it itself becomes a new kind of basic element. In turn, we might start asking what substances our theories, metaphors, and methods are based on.

PICK A SUBSTANCE: BROMINE

My encounter with this method began with a challenge to write about an element. I assumed that my lucky element would be the halogen fluorine: the basis of many of the PET brain scans I wrote about in *Picturing Personhood* (2004), the topic for a whole week in my Drugs, Science, and Culture class, and the subject of horrible conspiracies by aluminum companies, the military, the government, and dentists. But when I was given the challenge, I rolled the search engine dice, googling *element* and my current research project on fascia and connective tissue, and hit "I'm feeling lucky." The result was fluorine's big brother, bromine.

In 2014, bromine was declared the twenty-eighth essential element for animal life. Bromine (or Br) is a type of atom known as a halogen, full of high electronegativity that is highly reactive with other molecules. Fluorine, for example, "is so reactive that it is difficult to find a container in which it can be stored" (Bodner Research Web n.d.). Bromine is the only nonmetal that is a liquid at room temperature. A highly volatile and unstable compound, it is extremely reactive in order to obtain the extra electron it needs to be happy. According to Wadhawan (2016) on Quora, an atom "has its love or quest to take one or more electrons from another item which is in a position to give so as to fulfill its own configuration." Bromine, as we shall see, causes transformations—many of them toxic—in most places it is studied.

In what follows, I investigate bromine as a site for surprise in specialists. For each of my different research projects, I look for the bromine and then for those who find bromine so fascinating they want to see where it resists their understanding, who care for it so much that they want it to speak more than their discipline. In other words, I read the scientific literature and the internet more generally for bromine in the making, bromine escaping, "mondo" bromine (see Milburn 2015). Rather than reading for what is known, I am interested in how the unknown edges of elements are encountered, where they refuse to settle down and instead unsettle the concepts and methods of the researcher. This will look *very* technical at times. I do not understand all of the science.

My approach is to notice when the encounter between researchers and their substances generates palpable *affect*, places where they show their surprise, excitement, consternation, hesitation, confusion, and troubles, so much so that they *have* to say it in print. These are also markers of the durational *process* of doing science. Studies of the rhetoric of science have emphasized the peculiarly passive voice of "nature speaking" that constitutes the public and published discourse of facts being demonstrated (Haraway 1997; Latour 1987; Star 1989). Even though the succession of experiments is often not linear at all, a surprising result is usually retroactively reconstituted as having been the goal all along (Rheinberger 1997). In looking for sites of process and affect, I am attending to where researchers get stuck, where they realize that it is their own concepts that are part of the problem and have to invent new ways of talking and thinking in order to proceed. I think of this as the substances they are studying putting their vocabulary into variation. In noticing this, I am also learning how my own concepts might be just as limited.

CONNECTIVE TISSUE BROMINE

When bromine was declared the twenty-eighth essential element for animal life, it was briefly the center of science news, since, "despite its ubiquitous yet trace presence within animals, Br is without a known essential function" (McCall et al. 2014, 1381) (see figure 4.1). Bromine was immensely surprising to all the researchers involved when they found that it catalyzed the formation of a collagen scaffold (the basement membrane) that "anchors cells together to form tissues and organs" (Fidler et al. 2014, 331). Collagens are triple-helix elastic proteins that are mostly fibrillar (fiber-like) and can deform reversibly, playing structural roles throughout animals in between cells and in tendons, bones, and organs (Hynes 2009; Ricard-Blum 2011). They make up some 30 percent of proteins in mammals. This collagen-scaffold was termed a "primordial in-

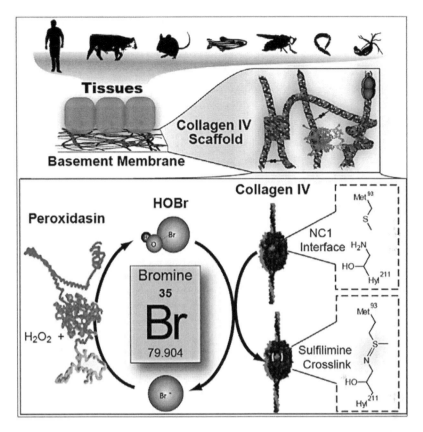

FIGURE 4.1: Graphical abstract that demonstrates the essential—but heretofore unknown and unlooked-for—catalytic role that bromine plays in the chemical reactions enabling collagen fibers to cross-link and form tissues that can support cell scaffolds, enabling multilayered life itself to evolve. From McCall et al. 2014.

novation" because it allowed the "emergence of larger, more complex animals able to resist predation and colonize new environments" (Fidler et al. 2014, 331).

For an element so critical to biology, it was surprising that bromine's role was only just discovered. The first reason is that cells were so much the focus of biology for the twentieth century that the things in between them were treated as secondary. With regard to tissue formation, chlorine could bond at the same places, and it was fifty thousand times more abundant in tissue areas than bromine. Researchers explained that they simply had not thought bromine could matter at such a low relative prevalence. This assumption became ironically self-fulfilling: "because Br has not been considered an essential trace element, systematic investigations on Br replacement had not been pursued" (McCall et al. 2014, 1390). Yet, bromine made all the difference in how the collagens bonded: it was a radical catalyst, overwhelmingly preferred over its more obvious rival; it enabled the bond and then it stepped out, to make more bonds. Its serial polyamory had been undetected, and affinity itself needed to be revised.

In this case, the fast and ultrarare had been overlooked; catalysis and reuse were revealed as more powerful than researchers had imagined possible. The cautionary tale for the scientists, and us, is how much science went along *just fine* without knowing this. This is the power and tragedy of theories: even as they make substances open to manipulation and understanding, they can simultaneously hide what might be really important.

As researchers figured out bromine's role, they also started experimenting to understand how this intercellular scaffold created functionality in the matrix of multicellular tissues. When they placed *Drosophila* (fruit flies) in a bromine-free environment, they found that the collagen IV scaffold functioned as "a 'molecular corset' to control egg shape" (McCall et al. 2014, 1387). This is a tale of misplaced passivity akin to Emily Martin's classic essay, "The Egg and the Sperm" (Martin 1991). As long as they had not looked to see how active collagen could be, they were content to attribute almost all the agency and control to cells. Whereas collagens had been known to provide structural strength, these researchers realized that they had discovered something fundamental: "It is plausible to argue that the evolution of multilayered organisms with their different cell layers separated by basement membranes was dependent on this basement membrane toolkit that has been maintained ever since" (Hynes and Naba 2012, 12). This compartmentalization of life within life enabled the evolution of multicellular organisms, as bodies with organs. We could say that bromine itself reimagined life.

Researchers continued to be surprised by how active this scaffold is: making cell migration and adhesion possible, enabling tissue compartmentalization, serving as a ligand for cell surface receptors, and patterning bone and tissue development, as well as providing structural integrity and mechanical signaling. "The cross-linked scaffold, in part, enables the evolution of complex tissues that overcome mechanical forces and the constraint of chemical diffusion in the delivery of nutrients to distant organs in large body size" (Fidler et al. 2014, 334). Here, the researcher (Fidler) echoes and builds on recent work on the *extracellular matrix* (ECM), the general term for "stuff" between cells (of which the basement membrane is a part).

The senior researcher on the bromine paper (McCall et al. 2014), Billy Hudson, had been obsessed with collagen since his early medical training working on kidneys where the membrane can malfunction. He moved to Vanderbilt in 2002 to start the Center for Matrix Biology, devoted to extracellular matrices (Snyder 2017). But seven years later, in 2009, an article could be published in *Science* still finding the ECM strange: "The Extracellular Matrix: Not Just Pretty Fibrils" (Hynes 2009).

The ECM is challenging for scientists because its metaphoric assumptions want it to be *one* thing—either architecture or signaling, either passive or active. For instance, Brown (2011, 4) writes, "ECM function in two ways. The first is an active or directive way, in which it imparts specific information to the cells that come in contact with it. The second is a passive or supportive role, in which it helps maintain the structure of tissues and the organism, without directing cellular behavior. It is not always easy to distinguish these roles, and many ECM structures are likely to contribute both functions." Here Brown struggles with his desire to keep his concepts distinct (active versus passive) while his observations demand a radically different way of understanding the ECM.

Other researchers develop a more process-oriented vocabulary. Schwartz (2010, 2) suggests, "It is obvious that the structures that bear the force should be involved in sensing it." Tissues for him are self-reinforcing processes: they sense force and can become stronger, thicker, stiffer in response to it (or the opposite). This turns out to be true of bones, artery walls, adhesions, and fascia that are constantly growing, shrinking, and adjusting in interdependence with surrounding tissues and forces. Where Schwartz is lured in by doubling his vocabulary, Brown (2011, 9–10) struggles with these apparent contradictions: "They are simultaneously alive and structural: This reflects one of the fascinating aspects of cellular and molecular architecture, the fact that most load-bearing structures still require turnover of the cells and molecules that make up the structure. Thus, the structures must retain their ability to function while old components are exchanged for new ones."

The ECM puts this researcher's vocabulary into variation most explicitly, generating both fascination and a tendency to oscillate: "It is still difficult to discern the balance between a role for the ECM in simply providing a permissive structural support or actively directing changes to cell and tissue morphology" (Brown 2011, 11). Thinking with dance researchers, I love this description: a balance between permissive structure and active direction is one definition of improvisation! Perhaps we can think of the ECM as improvisatory and, in turn, start thinking of improvisation in tissue-like terms.

Pressure on a basement membrane is met with growth. Even small impacts are met by thickening walls, strengthening adhesions, and so on. What seems to be a barrier or structure is actively shaping its environment and reshaping itself. One lesson for my theories and concepts is to look for the work and activity in what I tend to treat as relatively fixed structures—to see that they are processes, structuring. Structuring is a process of ongoing maintenance and responsiveness, lively and even anticipatory.

Watching the progress of researchers such as Brown who were surprised by the things they cared about, I sense in their struggle a recognition of the limits imposed on their research because of what they think their substances can do. Once I started looking for it, I realized that what the scientists often said was that the substances they were empirically investigating revealed how limited their theories were—precisely because those theories had been modeled on different substances (and their elements were modeled on different elements). This was my introduction to the idea that a substance can become a method of undoing one's theoretical assumptions. The refusal of the substance to conform to existing binaries in a researcher's training (e.g., alive versus structural) pushes researchers to think and measure differently, to undo their current practices of using instruments and to explore new methods.

LOOKING EVEN MORE AT FASCIA

The primary tissue that collagen makes in bodies, connective tissue in all of its variations, is often called fascia. It can be said to wrap every "thing" in the body, muscles but also individual muscle fibers, nerves, arterioles, organs, parts of organs, and so on. It divides and connects; it is the way parts relate to each other; it is relation (Dumit and O'Connor 2016). Japp van der Wal (2015, 103) calls fascia "middleness" and "innerness." The International Fascia Research Congress is where many who work and live with and love fascia—fasciaphiliacs—have to create new languages and metaphors to handle it as it handled them: whether biologists, clinicians, massage therapists, dissectors, or dancers. But they constantly came up against the limits of their existing tools and vocabulary for accounting for their observations and for doing work together. So they formed a committee specifically to try and create a new language: "the FNC [Fascia Nomenclature Committee] recognised that conventional (anatomical) fascia-relating language lacks the linguistic capacity required to effectively discuss fascia's morphology, architectural distribution, material properties, and physiological roles (e.g., mechanical force transmission, sensory capacity) from a more holistic ideological position" (Adstrum et al. 2017, 174).

The committee turned to the concept of *ideology* in order to grapple with the challenge posed by empirical observations. Having to come up with new grammars makes them aware of the conventionality of their existing concepts and the limitations of their foundational frames of reference. Holism is, by default, a larger standpoint, serving the function of allowing *more* into the frame. Similarly, another researcher stated, "To understand how collagens work in a concerted fashion with their extracellular and cell-surface partners, a global, integrative approach is needed" (Ricard-Blum 2011, 15). Collagens themselves are, like cells, still in the throes of metaphoric struggles: despite being the most abundant proteins in mammals, this review stated that "even the question 'What is collagen, what is not?' may still be valid" (1).

One main impetus for scientists realizing how much more there might be to find in collagens, scaffolds, membranes, and the extracellular matrix comes from the stunning endoscopic video work by plastic surgeon Jean-Claude Guimberteau. Because of his early investigations into living tissue, by 2005 he had access to contact endoscopy with high-definition resolution, enabling full-color, highly magnified videos (30X up to 400X) in real time inside a living body, showing fibrils, individual cells, and bacteria in action. These videos, with titles such as *Strolling under the Skin* (2003) and *Muscle Attitudes* (2010), showed that fascia is unbelievably fractal and that the living matter of bodies is a unified whole, in which all available space is occupied by viscous structures in dynamic tension with each other. When Guimberteau showed *Strolling* at the First Fascia Congress in 2007, it was a revelation to almost everyone, even those who had been investigating fascia for years. The videos transformed their intuitions into full demonstrations of the strange, living biotensegrity of viscous connective tissue (and if I could include a video in this chapter, it would be *Strolling*).

Guimberteau discovered for himself and then showed others that "when you place an endoscope with a camera inside the living tissue of a patient, your perception changes, and the underlying assumptions and 'generally accepted truths' [of medicine and surgery that he had been trained in] underpinning these laboratory-based hypotheses no longer seem as reliable" (Guimberteau and Armstrong 2015, 5). He was so surprised by what he saw that he reoriented his life to serve this fascination:

> The description of these structures in the classic anatomy textbooks provided a reassuring and logical theory of the movement of tendons within their sheaths, but I found this to be completely inaccurate and hopelessly inadequate when I started observing living tissue through an endoscope.
>
> I therefore began to pay close attention to the study of this connective tissue, which has long been neglected by surgeons and anatomists. I was surprised to discover that it is composed of a network of collagen fibers, which are arranged in a completely disorderly fashion with no apparent logic. I could have abandoned the task of trying to understand this complex organization of this tissue, but I was intrigued by the fact it seemed to ensure the efficient, independent movement of adjacent structures with great precision and finesse. Could apparent chaos and efficiency coexist? (Guimberteau and Armstrong 2015, 10)

Collagen's challenge to his assumed opposition between categories of chaos and efficiency leads to his becoming obsessed with this substance. He must pay due attention to it, even when it demands more than he is capable of saying. When it suggests a paradox, he becomes excited by the possibility that his training and his methods have been deeply wrong. Agency is at stake, since he finds the substance doing things that seem to go against its nature as a passive object. Yet this researcher is unable not to follow this intriguing substance and can no longer conform it to now-seemingly inane strictures (of opposites, of words, of formulations of agencies, boundaries, parts).

Guimberteau turns this accusation around, insisting on staying in closer contact with his substance. He comes to realize that "it is important to avoid being seduced by theories that may be conceptually attractive but that in reality are inaccurate. Therefore we will first describe what we can observe through an endoscope, then we will try to make sense of what we have seen" (Guimberteau and Armstrong 2015, 4). This is a type of radical empiricism that nonetheless knows it is inside of history, since if we are only just discovering how limited our view has been, there is every expectation that these observations are also theory-laden, attractive but inaccurate. "Future generations of anatomists will probably smile at these discoveries" (109). He does not know if he is progressing, only that he cannot stop looking.

If we pay attention in this manner, we realize that each substance resists the work and curiosity of its specialist in its own way; each requires extensive exploration to understand its properties. Often, different or new tools are needed to figure out what it does, how it relates, connects, or does things, and how it refuses to do other things. In particular, the verbs that have been used to describe it hold back speculation and themselves need empirical work. The specialist often comes to realize that the substance might have new types (e.g., a pro tennis player starts to think about different types of wood as very different substances) or might behave totally differently in different environments or milieus or when near or connected to certain other substances. Guimberteau's work on fascia challenges me to think about how my concepts may conceal in-betweens that I take for granted and to ask whether those in-betweens may be as agentive, full, and improvisational as what surrounds them.

PHARMACEUTICAL BROMINE

Wondering what other surprises might come from bromine, I explored its role in pharmaceutical design, the subject of my second book. As I looked at recent research into halogen bonding, it turned out that bromine and other halogens had been shaping molecules in ways that had been as discounted as bromine in tissues. "For some time, the significance of halogen bonding to biological macromolecular structure was overlooked . . . primarily because halogens are not commonly found in naturally occurring proteins or nucleic acids" (Ho 2014, 244). It took a 2004 survey of the Protein Data Base (PDB), where all newly mapped protein configurations are registered, to reveal just how important halogen bonds had become in creating ligands for use in biology and medicine.

Halogen bonds were not new to the chemistry community, but they were certainly strange to everyone who researched them. By 1968, one reviewer could "list some 20 descriptive phrases used during the 'first century' of XB [halogen bonds], illustrating the struggle to understand the phenomenon. They varied from the imaginative 'bumps in hollow' to the chemically meaningful 'charge-transfer interaction'" (Cavallo et al. 2014, 12). It was weird and surprising to find that halogens put the whole notion of bonding into variation. And so a whole new type of bond needed to be named: X-bonds.

A halogen bond is doubly strange: while the bromine in tissue making is covalent (the type of atoms stuck together in your typical ball-and-stick models that share electrons), the halogen bond is *noncovalent*, with atoms interacting at a distance. It functions *like* hydrogen bonding—one of the things that gives water its magical, nonlinear properties, often drawn as a dotted line that really needs three dimensions to appreciate, sharing and not sharing atomic space. But halogen bonds put hydrogen bonding into variation, deforming the imagined sphere of the atoms and, in the process, deforming the molecule. Textbooks have been forced to comment on what Karen Barad (2011) would call the queerness of it all. Halogens love too much, and too strangely: "Atoms in molecules have been approximated, in the past, as interpenetrating spheres and halogen atoms have long been considered neutral spheres in dihalogens and negative spheres in halocarbons. . . . Electrophilic halogens *thus appeared strange*, and *the persistent biases* resulting from the approximations described above long prevented electrophilic halogens from being recognized as responsible for the formation of relatively strong and highly directional interactions in the solid, liquid, and gas phases" (Cavallo et al. 2014, 7; emphasis added).

After initial attention was paid to bromine bonds in proteins, other researchers responded first by ignoring them, then by treating them as if they were hydrogen bonds, and only grudgingly by coming to see them as not only different but surprisingly so and thus in need of a grammar of their own. Even when they were first seen, they could not be understood as such. "The simplistic understanding of halogens by us, as biologists, was at odds with the observation that this electron-rich atom formed such a close interaction with a formally negatively charged oxygen, and that it would replace a stabilizing H-bond—the X-bond concept was absent" (Ho 2014, 248). The biologists here came to see themselves as captive to concepts based on another element. They thought they could treat halogens as if they had hydrogen-like bonds. Therefore they did not notice the new type of transformation that occurred with X-bonding.

Once they were granted this new name, X-bonds in turn placed *biological* systems into variation, and the result needed to be called "biomolecular X-bonds (BXBs) to emphasize their distinctive nature" (Ho 2014, 244). In pharmaceutical design, when a scientist uses bromine to replace a hydrogen atom in a protein, the BXB it forms warps the protein just enough to change its function, creating a new protein that does the right thing (binds to the same enzyme) and the wrong thing (blocks its activity, decreasing its function) at the same time: it becomes a possible inhibitor.

The relative simplicity of creating potential inhibitors should have easily led to new pharmaceutical design. But the innovation did not happen, because ignorance of bromine had been embedded ideologically in the software that chemists used to model and predict proteins. Ho describes how computational pharmaceutical chemists start looking for molecules in large repositories that are available for testing and then "perform large scale virtual docking studies as a first screen for compounds. [But] the standard docking algorithms, with their preprogrammed properties for halogens, would not only be incapable of modeling an X-bond, they would in fact have the halogen and its Lewis base [binding] partner repelling each other. . . . In short, 20% of potential lead compounds could be discarded in the first screen simply because the current docking programs do not treat these bonds properly" (Ho 2014, 245).

In other words, the algorithms the chemists were using to model halogens were built with theories based on different elements, and in turn the chemists missed key drugs, *by design*. Now that X-bonds have been properly encoded into the software, halogenated molecules represent a large proportion of the pharmacological compounds currently in clinical use or in clinical trials. They also constitute a significant share of the compounds I have been studying in the pharmaceutical industry.

We can, therefore, trace the way that these researchers, caring for the specificity of their current substances, discovered the prior substances that their existing theories were absolutely dependent on. They were turned on to a truly new substance at the moment when their world was undone and redone. This is love—and maybe a reorientation toward materiality and a "new appreciation of possibilities inherent in our relation to world" (Mathews 2005, 7). For the biologists, then, this became their religion: "For the medicinal chemist, BXBs are now a regular if not common tool in the molecular design toolbox. There remain, however, many audiences in which the question of 'Who has heard of halogen bonds?' is met with deadly silence. Thus, there is still work to be done for the evangelist to spread the word of this molecular interaction which has been long neglected, but which has great potential in the biological arena" (Ho 2014, 270).

As much as specific substances can spawn new theories, new software, and new products for these researchers, we can also note how all of our theories are in correspondence with often implicit substances: writing back and forth with some substances and not others. Covalent bonds make noncovalent bonds look abnormal. And noncovalent hydrogen bonds kept researchers from noticing that X-bonds were something altogether different.

In this correspondence, we approach something that Hans Blumenberg calls "absolute metaphors" (2016). In his *Paradigms for a Metaphorology*, he describes a method for understanding the way in which we are full of concepts (such as "truth") that depend absolutely on metaphors, because they are not empirical in any simple sense. Therefore, they can be approached only via metaphors that possess them absolutely and that shape our understanding in advance, always already, in something akin to ideology, paradigm, or epoch.

TOXIC BROMINE

Treating each substance as a method opens us up to surprise about our own methodological assumptions and metaphoric dependencies. Fascia challenged me to rethink structure and in-betweenness as a potentially ongoing improvisation. Halogen bonding challenged me to rethink relations as generative and to consider how they might be contingent on the instruments used to reveal them. In the final section of this chapter, I explore bromine's surprising toxicity. Again, my goal is not to be comprehensive but simply to look into those sites where the substance caused researchers who loved it to rethink their own approaches.

By the end of the twentieth century, bromine was known to pose a serious danger to public health, but there were many assumptions—not to mention willful and industrially designed ignorance—about its ability to transform bodies and travel environmentally (Stockholm Convention on Persistent Organic Pollutants 2015) (see figure 4.2). For instance, when lead was first added to gasoline to improve engine performance, it was found that deposits built up, eventually clogging the engine and creating knocking sounds. The solution was to mix brominated chemicals with the gasoline. As the fuel burned, the bromine combined with the lead to produce lead bromide, which readily passed out through the exhaust. The result was that a poisonous heavy metal (lead) was dispersed throughout our cities (Knight 2014).

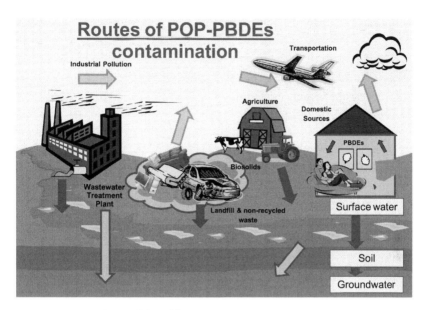

Routes of POP-PBDEs contamination

Industrial Pollution

Transportation

Agriculture

Domestic Sources

PBDEs

Biosolids

Wastewater Treatment Plant

Landfill & non-recycled waste

Surface water

Soil

Groundwater

FIGURE 4.2: This chart is labeled "Routes of POP-PBDEs contamination," and yet it and others like it extract the molecules from the sociogenic unequal distribution of life chances and so are equally a mystification of the alterlife we are becoming. Inside the body, PCBs and PBBs do things precisely because they bind in ways that mess with the proteins they mimic, producing hormonal or antihormonal effects. These are precisely the qualities that make Br effective in drug design. From Stockholm Convention on Persistent Organic Pollutants 2015.

But bromine's main use for the last forty years has been as a flame retardant. It is in the computer I am using to write this article, in every television and most other electronics, and, for many years, in all of our couches and blankets. When a flame causes combustion, free radicals (highly oxidizing agents) are produced, which "are essential elements for the flame to propagate. The extreme affinity halogens have made them very effective in trapping free radicals, hence removing the capability of the flame to propagate" (Guerra et al. 2010, 4). Bromine is so hyper-reactive that it is very effective at stopping fires by grabbing all the oxygen (Lauren Knight, "Who's Afraid of Bromine?," BBC News Magazine, September 27, 2014).

If the fire does not stop, however (or if the electronics are melted down for recycling), then the heat consumes the "retardant" material, and the bromine starts combining with many other things and is released into the atmosphere. This process produces PBBs, PBDEs, PBDDs (polybrominated biphenyls, polybrominated diphenyl ethers, polybrominated dibenzo-p-dioxins), and other congeners. Though we do not know the full extent of their toxicity, PBDDs are similar enough to the terrifying chlorinated dioxins to have been declared "among the most toxic byproducts" (Zhang, Buekens, and Li 2016, 27). These are close cousins of the PCBs and PCBE that first caused panic and scientific investigation when they were discovered to be rising in breast milk in 1998. The discovery of flame-retardant toxicity "astounded the scientific community" (Guerra et al. 2010, 10; see also Meironyte, Noren, and Bergman 1999) since it had not worked through the multispecies transport of these molecules. "Persistent organic pollutants (POPs), such as brominated flame retardants (BFRs), are organic compounds that are resistant to environmental degradation through chemical, biological, and photolytic processes. Because of this, they have been observed to persist in the environment, to be capable of long-range transport, to bioaccumulate in human and animal tissue, to biomagnify in food chains, and to have potential significant impacts on human health and the environment. However, there are a number of reactions that can take place once the BFRs are released to the different environmental compartments (air, water, soil, sediment)" (Eljarrat, Feo, and Barceló 2011, 189).

Some of the worst outcomes from fires in towns result from the toxins released from the retardant byproducts. In 2012, a powerful series of investigative journalism articles in the *Chicago Tribune*, "Playing with Fire," described how attempts to ban flame retardants were met by fierce opposition that delayed regulations for decades (Callahan and Roe 2012a). Although the resistance seemed to be made up of grassroots activists for safety and firefighters, in fact they were flame-retardant manufacturer groups with names such as "Citizens for Fire Safety" and "the Bromine Science and Environmental Forum." These "astroturf" (fake grassroots) groups explicitly claimed that members of minority groups would be most at risk from the proposed ban. Helmed by former tobacco-industry marketers, these organizations constituted the second wave of probromination organizations (Callahan and Roe 2012b). The first wave started as early as the 1970s with tobacco companies fighting against the fears that cigarettes started fires. Through lobbying efforts and the corruption of fire marshal organizations, they tried to prevent fire-safe cigarette standards by focusing attention "on the fuels rather than ignition sources" (Callahan and Roe 2012b, quoting a 1996 strategic plan by R. J. Reynolds tobacco company). And they succeeded. But when cigarettes were finally required to be fire-safe, the tobacco companies dropped their monetary support of fire marshal organizations, but the flame-retardant companies picked it up and further amplified the use of flame retardants. All this was done even as studies showed that the retardants did not prevent fires and in fact made smoke more toxic.

Even worse, the bromine flame-retardant chemicals are only imperfectly bonded to the materials they are supposed to protect, and only imperfectly bonded to each other. Although reviews comment on the nascent state of the field, bromine compounds turn out to be incredibly agile at multiplying their forms and their forms of travel. Some, including PBDEs, "migrate comparatively readily"—like dust—from such textiles as furniture and baby sleepwear into the air, skin, and mouth. Others are more bonded and therefore, for instance, "there are low concentrations of TBBP-A reported in indoor air, dust and food" (Stubbings and Harrad 2014, 167).

Even though the manufacture of new brominated fire retardants (BFRs) was banned in the 2000s, the textiles and electronics are still omnipresent in homes across the world and make up a large amount of landfills. Landfills themselves are the subject of fascination for many biologists and chemists, who struggle to model the varied layers and processes taking place within them over long periods of time. The more these scientists look, the more they find—including that the substances and bacteria in landfills go through different metabolic stages, including aerobic, anaerobic, and methanogenic bacterial metabolism. Each phase privileges a wildly different set of transformative potentials. For BFRs, the result is that they can change forms many times. Even so-called less persistent and bioaccumulative and potentially less toxic ones such as BDE-209 can be debrominated into PBDEs that persist and bioaccumulate, as well as becoming biotransformed into dioxins and other toxins (Stubbings and Harrad 2014).

We have learned so little about these substances, partly because they have not been investigated, and partly because they are so prolific. Researchers describe being overwhelmed and fatigued by the very attempt to study the variations of potentially (and probably) toxic products (cogeners) of these metabolic transformations. Zhang, Buekens, and Li (2016, 27) note that "the sheer number of PXDD/Fs congeners [variations of substances] precludes systematic analysis of individual congeners" and estimate that there are close to nine thousand different ones. Those that are suspected but unstudied are therefore technically not yet regulatable as toxic. Additionally, as with leaded gasoline, the presence of bromine or hydrogen bromide in fires containing other heavy metals leads to their volatilization into the atmosphere. Affectively, the apparently exciting multiplicity of variations for BFRs has reached a scientific stress point: "The complexity of chemical analysis and the sequence of phenomena render the study and analysis of the rare experimental results tedious and difficult" (36). Even though they are toxic, they may be too tiring and too boring to research. Tragically, under current US regulations, boredom and fatigue occur because the research increases costs for governments rather than making money for companies.

A final form of surprising bromine toxicity relates to another of my preexisting research topics, the effects of fracking on drinking water. In oil and natural gas wells, solutions containing zinc bromide are used to displace drilling mud, and this substance's heaviness makes it useful in holding back flammable oil and gas particles in high-pressure wells. Zinc bromide is so inexpensive that drilling fluid makes up 16 percent of the bromine market in the United States (Reisch 2015). But when that fluid gets miles under the ground, it helps break up (fracture) the earth, resulting in the freeing of a grand mix of minerals, including yet more bromide. These come back up as contaminated "flowback fluid" or, as the industry calls it, "produced water" (Arnaud 2015; see also Gregory, Vidic, and Dzombak 2011).

In one of those magical terms that lends credence to the corporate construction of reality, all of this fracking fluid and "produced water" is legally regulated as water and therefore not subject to the jurisdiction of the Environmental Protection Agency, but it is also excluded from the Pure Water Drinking Act, so it is not regulated as water either (Wylie 2018). Legally, it simply does not exist. For years, some of it was sent into wastewater treatment plants to be turned into drinking water. But it contains bromide, and bromide places wastewater treatment into variation: turning the very process of purification into its opposite. One of the main techniques of making water safer is by disinfecting it, killing the microbes through chlorination, chloramination, or ozone. But these processes actually react with even a little bit of bromide to create *bromate*, a potent carcinogen, and many other problematic and toxic compounds. This process results in a so-called disinfectant byproduct— which in noneuphemistic language would be called an infectant. We can see here an opportunity to continue to rethink production through by-production (Parker et al. 2014).

Coca-Cola, in fact, rediscovered this deadly process in the United Kingdom when it was trying to launch its bottled water, Dasani, which was created from treated tap water from a city on the River Thames. The company constructed its own reality with what it called a "highly sophisticated purification process" (BBC News 2004). "Add a batch of calcium chloride, containing bromide, for 'taste profile'; then pump ozone through it, oxidizing the bromide —which is not a problem—into bromate —which is" (Felicity Lawrence, "Things Get Worse with Coke: Bottled Tap Water Withdrawn after Cancer Scare," *Guardian*, March 19, 2004). This led to a massive recall of Dasani in the United Kingdom and destruction of the brand throughout Europe.

Bromide, in the case of water treatment plants, resists the apparent stability of "treatment." Pennsylvania finally banned fracking fluids in its treatment plants in 2011, and hundreds of millions of gallons of fracking flowback fluid was shipped to Ohio and injected underground, where it led to earthquake swarms. And even when banned, 14 percent of wells have violated existing policies regarding disposal of fracking brines, and 30 percent of major reported spills are due to brine spills (Parker et al. 2014). Substances create their own language and metaphors, though not ones that help save the world. Corporations can easily use the wiliness of substances, their queerness and nontranslatability, as a form of refusal to know, of denial and suppression.

In each of these cases of toxicity, bromine demands its own form of life-cycle analysis. The toxic worlds we live in persist in part because the main forms of analysis we have are not only wrong but misleadingly so (and easily exploited by those who would prefer we know less about toxins). Bromine continues to be used in part because it is so cheap: it is hard to get rid of because it is baked into our economy, which sees expense as fungible with health. It produces unhealthy byproducts: many inequalities of health and death along the way. We too can consider how limited our analyses of life cycles of products and breakdown are and how many assumptions we make about what our substances produce and what they become. Alongside the researchers described above, we too might look to the areas where we feel overwhelmed by multiplicity and examine whether social justice demands that we face the tedious and difficult work of tracing them out.

There is so much more to all of these stories. But perhaps I can leave with you some of the lessons I have learned in this process. These are not new ideas; they are present already in the anthropology of science, medicine, and science and technology studies. I have been using books such as *Testo Junkie* (Preciado 2013), *Gut Feminism* (Wilson 2015), and *Mushroom at the End of the World* (Tsing 2017) to teach classes using substances as explicit methods and pedagogies.

Testo Junkie follows testosterone as it is historically manufactured, isolated, and named, as it creates an industry and a norm, and as it is shaped by norms and industries. The book examines the capacity of others to select it, use it to biodrag, and render parts of worlds manipulable in different ways that are *unique* to testosterone in selecting, interpreting, modifying, and producing gendered "bodies," as separate from minds. Preciado's method of testosterone works in a manner that is *almost* the inverse of the way in which antidepressants produce, modify, and select "mind" separate from body in Wilson's *Gut Feminism*. Each is specific. Bromine "binds," "travels," and "transforms" in its own fashion. Each substance opens up its own selections and productions, its own substantializations, its own elemental units, its own worlds.

If each substance challenges the words and worlds we can use to talk about what it modifies and relates to, then each messes with parts of our theoretical vocabulary, our assumptions of reality and malleability and relationality. Therefore, we can play with turning this around and ask of our theories: what substances make them up? How are we taking them as elemental, and how lively might those elements turn out to be? The conceptual tools that we use to approach a substance may themselves depend on specific substances and notions of substances. In other words, part of the challenge of thinking with substances is that each substance puts into question the very words/concepts/things/practices/methods we depend upon.

So I offer the following as a playful mode of inquiry, as my way of thinking about methods. It is one that may open up new possibilities but may not—it is not a prescription (see Dumit 2021). Pick a substance related to (but not central to) what you are studying, and do some research. Find people for whom it might be a fetish. Locate some specialists who care about that substance, and read some of their work (scientific papers, treatises about working with that substance, etc.). This is where you locate the words and verbs that others have had to invent or to give new meaning to, in order to deal with the specificity of the substance. These are the verbs and properties that may help you think and write better. These put the binaries into variation: each substance probably generates a different notion of matter and a different notion of form in a different kind of relation to other substances.

Do not just play with them etymologically or symbolically: this stays within your theory, within your paradigms. It is incredibly productive of words but rarely jostles your being. I also suggest that you avoid a substance central to your research, because you think you already know it. It is the proper subject of an "implosion exercise" (Dumit 2014), in that you map the substance in the world and the world in the substance, but here you are looking for the places where a substance demands its own world.

Follow the substance out to the specialists who have to live with it, who love and obsess about it, who have an abnormal appetite or liking for it. Talk to the philiacs—the scientists, yes, but also lovers, gardeners, craftspeople, artists, engineers, cooks, technicians, athletes, workers, and so on. The writing may be technical, but you can scan to find out where the specialists are surprised by their substance. Here they break ranks with the passive voice of research and cannot help but speak their own hesitations, confusions, and excitement.

In their grappling with the substance in their lives, they adapt with it and invent ways of being with it, and you can find where they have had to invent theories to account for their interactions and vocabularies specific to the substance. Look at where they run into problems and talk about the limits of their tools for thinking and making. This is where they enable a substance to become its own metaphor and undo previously absolute metaphors of their existence and relations. Those moments of surprise, improvisation, and invention are also helpful for you as a writer and theorist to see in your own thoughts and theories their substantial limitations. You can then trace these back to the genre of the human (Wynter 2003) who is asking the questions.

The results of playing with substance as method are not answers but new forms of questions—new methods, perhaps, but not necessarily better ones, of course. Above all, each substance resists being lumped into matter in general. Each demands its own materiality.

REFERENCES

Adstrum, Sue, Gil Hedley, Robert Schleip, Carla Stecco, and Can A. Yucesoy. 2017. "Defining the Fascial System." *Journal of Bodywork and Movement Therapies* 21 (1): 173–77. https://doi.org/10.1016/j.jbmt.2016.11.003.

Arnaud, Celia Henry. 2015. "Figuring Out Fracking Wastewater." *Chemical and Engineering News*, March 16. https://cen.acs.org/articles/93/i11/Figuring -Fracking-Wastewater.html?type=paidArticleContent.

Barad, Karen. 2011. "Nature's Queer Performativity." *Qui Parle: Critical Humanities and Social Sciences* 19 (2): 121–58.

BBC News. 2004. "Coke Recalls Controversial Water." March 19. http://news.bbc .co.uk/2/hi/business/3550063.stm.

Blumenberg, Hans. 2016. *Paradigms for a Metaphorology.* Ithaca, NY: Cornell University Press.

Bodner Research Web. n.d. "The Chemistry of the Halogens." Accessed January 4, 2018. http://chemed.chem.purdue.edu/genchem/topicreview/bp/ch10/group7 .php.

Brown, N. H. 2011. "Extracellular Matrix in Development: Insights from Mechanisms Conserved between Invertebrates and Vertebrates." *Cold Spring Harbor Perspectives in Biology* 3 (12): a005082. https://doi.org/10.1101/cshperspect .a005082.

Callahan, Patricia, and Sam Roe. 2012a. "Playing with Fire." *Chicago Tribune.* http://media.apps.chicagotribune.com/flames/index.html.

Callahan, Patricia, and Sam Roe. 2012b. "Big Tobacco Wins Fire Marshals as Allies in Flame Retardant Push." *Chicago Tribune.* May 8. https://www.chicagotribune .com/lifestyles/health/ct-met-flames-tobacco-20120508-story.html.

Cavallo, Gabriella, Pierangelo Metrangolo, Tullio Pilati, Giuseppe Resnati, and Giancarlo Terraneo. 2014. "Halogen Bond: A Long Overlooked Interaction." In *Halogen Bonding 1: Impact on Materials Chemistry and Life Sciences*, edited by Pierangelo Metrangolo and Giuseppe Resnati, 358:1–17. Cham, Switzerland: Springer. https://doi.org/10.1007/128_2014_573.

Dumit, Joseph. 2004. *Picturing Personhood: Brain Scans and Biomedical Identity.* Princeton, NJ: Princeton University Press.

Dumit, Joseph. 2014. "Writing the Implosion: Teaching the World One Thing at a Time." *Cultural Anthropology* 29 (2): 344–62. https://doi.org/10.14506 /ca29.2.09.

Dumit, Joseph. 2021. "Substance as Method." In *Experimenting with Ethnography*, ed. Andrea Ballestero and Brit Ross Winthereik. Durham, NC: Duke University Press.

Dumit, Joseph, and Kevin O'Connor. 2016. "Sciences and Senses of Fascia: A Practice as Research Investigation." In *Sentient Performativities of Embodiment: Thinking alongside the Human*, ed. L. Hunter, E. Krimmer, and P. Lichtenfels, 35–54. Lanham, MD: Lexington Books.

Eljarrat, E., M. L. Feo, and D. Barceló. 2011. "Degradation of Brominated Flame

Retardants." In *Brominated Flame Retardants*, edited by Ethel Eljarrat and Damià Barceló, 187–202. Heidelberg, Germany: Springer. https://doi.org/10.1007/698_2010_96.

Fidler, A. L., R. M. Vanacore, S. V. Chetyrkin, V. K. Pedchenko, G. Bhave, V. P. Yin, C. L. Stothers, et al. 2014. "A Unique Covalent Bond in Basement Membrane Is a Primordial Innovation for Tissue Evolution." *Proceedings of the National Academy of Sciences* 111 (1): 331–36. https://doi.org/10.1073/pnas.1318499111.

Gregory, K. B., R. D. Vidic, and D. A. Dzombak. 2011. "Water Management Challenges Associated with the Production of Shale Gas by Hydraulic Fracturing." *Elements* 7 (3): 181–86. https://doi.org/10.2113/gselements.7.3.181.

Guerra, P., M. Alaee, E. Eljarrat, and D. Barceló. 2011. "Introduction to Brominated Flame Retardants: Commercial Products, Applications, and Physicochemical Properties." In *Brominated Flame Retardants*, ed. E. Eljarrat and D. Barceló, 16:1–17. New York: Springer Berlin Heidelberg. https://doi.org/10.1007/698_2010_93.

Guimberteau, J. C. 2003. *Strolling under the Skin*. Video. Pessac, France: ADF Video Productions. https://www.youtube.com/watch?v=eWolvOVKDxE.

Guimberteau, J. C. 2010. *Muscle Attitudes*. Video. https://www.youtube.com/watch?v=QjMpaRfjOmc.

Guimberteau, J. C., and Colin Armstrong. 2015. *Architecture of Human Living Fascia: The Extracellular Matrix and Cells Revealed through Endoscopy*. Edinburgh: Handspring.

Haraway, Donna J. 1997. *Modest_Witness@Second_Millennium.FemaleMan_Meets_OncoMouse: Feminism and Technoscience*. New York: Routledge.

Ho, P. Shing. 2014. "Biomolecular Halogen Bonds." In *Halogen Bonding 1: Impact on Materials Chemistry and Life Sciences*, edited by Pierangelo Metrangolo and Giuseppe Resnati, 358:241–76. Cham, Switzerland: Springer. https://doi.org/10.1007/128_2014_551.

Hynes, R. O. 2009. "The Extracellular Matrix: Not Just Pretty Fibrils." *Science* 326 (5957): 1216–19. https://doi.org/10.1126/science.1176009.

Hynes, R. O., and Naba, A. 2012. "Overview of the Matrisome—An Inventory of Extracellular Matrix Constituents and Functions." *Cold Spring Harbor Perspectives in Biology*, 4 (1): a004903. https://doi.org/10.1101/cshperspect.a004903.

Latour, Bruno. 1987. *Science in Action: How to Follow Scientists and Engineers through Society*. Cambridge, MA: Harvard University Press.

Martin, E., 1991. "The Egg and the Sperm: How Science Has Constructed a Romance Based on Stereotypical Male-Female Roles." *Signs: Journal of Women in Culture and Society* 16 (3): 485–501.

Mathews, Freya. 2005. *Reinhabiting Reality: Towards a Recovery of Culture*. Albany: SUNY Press, 2005.

McCall, A. Scott, Christopher F. Cummings, Gautam Bhave, Roberto Vanacore, Andrea Page-McCaw, and Billy G. Hudson. 2014. "Bromine Is an Essential Trace Element for Assembly of Collagen IV Scaffolds in Tissue Development and Architecture." *Cell* 157 (6): 1380–92. https://doi.org/10.1016/j.cell.2014.05.009.

Meironyte, Daiva, Koidu Noren, and Ake Bergman. 1999. "Analysis of Polybrominated Diphenyl Ethers in Swedish Human Milk: A Time-Related Trend Study, 1972–1997." *Journal of Toxicology and Environmental Health*, Part A 58 (6): 329–41. https://doi.org/10.1080/009841099157197.

Milburn, Colin. 2015. *Mondo Nano: Fun and Games in the World of Digital Matter.* Durham, NC: Duke University Press.

Parker, Kimberly M., Teng Zeng, Jennifer Harkness, Avner Vengosh, and William A. Mitch. 2014. "Enhanced Formation of Disinfection Byproducts in Shale Gas Wastewater-Impacted Drinking Water Supplies." *Environmental Science and Technology* 48 (19): 11161–69. https://doi.org/10.1021/es5028184.

Preciado, Paul B. 2013. *Testo Junkie: Sex, Drugs, and Biopolitics in the Pharmacopornographic Era.* New York: Feminist Press at the City University of New York.

Reisch, Marc S. 2015. "Bromine Comes to the Rescue for Mercury Power Plant Emissions." *Chemical and Engineering News* 93 (11): 17–19.

Rheinberger, Hans-Jörg. 1997. *Toward a History of Epistemic Things: Synthesizing Proteins in the Test Tube.* Stanford, CA: Stanford University Press.

Ricard-Blum, S. 2011. "The Collagen Family." *Cold Spring Harbor Perspectives in Biology* 3 (1): a004978. https://doi.org/10.1101/cshperspect.a004978.

Schwartz, M. A. 2010. "Integrins and Extracellular Matrix in Mechanotransduction." *Cold Spring Harbor Perspectives in Biology* 2 (12): a005066. https://doi.org/10.1101/cshperspect.a005066.

Snyder, Bill. 2007. "VUMC Reporter Profile—Billy Hudson's Ambitious Math, Science Education Plan Honors His Rural Roots (04/20/07)." Accessed January 4, 2018. http://www.mc.vanderbilt.edu/reporter/index.html?ID=5485.

Star, S. L. 1989. *Regions of the Mind: Brain Research and the Quest for Scientific Certainty.* Stanford, CA: Stanford University Press.

Stengers, Isabelle. 2011. *Thinking with Whitehead: A Free and Wild Creation of Concepts.* Cambridge, MA: Harvard University Press.

Stockholm Convention on Persistent Organic Pollutants. 2015. "Revised Draft Guidance for the Inventory of Polybrominated Diphenyl Ethers under the Stockholm Convention." Conference of the Parties to the Stockholm Convention on Persistent Organic Pollutants, seventh meeting, Geneva, May 4–15. http://chm.pops.int/Implementation/NIPs/Guidance/GuidancefortheinventoryofPBDEs/tabid/3171/Default.aspx.

Stubbings, William A., and Stuart Harrad. 2014. "Extent and Mechanisms of Brominated Flame Retardant Emissions from Waste Soft Furnishings and Fabrics: A Critical Review." *Environment International* 71 (October): 164–75. https://doi.org/10.1016/j.envint.2014.06.007.

Tsing, Anna Lowenhaupt. 2017. *The Mushroom at the End of the World: On the Possibility of Life in Capitalist Ruins.* Princeton, NJ: Princeton University Press.

Van der Wal, Japp. 2015. "Fasciasophy—Philosophical Aspects of an Organ of

Innerness." Powerpoint presentation on http://www.embryo.nl. Accessed April 1, 2021.

Wadhawan, Narinder. 2016. "What Is Meant by Electron Affinity?" Answered December 18 on Quora. Accessed May 17, 2017. https://www.quora.com /What-is-meant-by-electron-affinity.

Whitehead, Alfred North. (1920). 2015. *The Concept of Nature: Tarner Lectures*. Cambridge: Cambridge University Press.

Wilson, Elizabeth A. 2015. *Gut Feminism*. Durham, NC: Duke University Press.

Wylie, Sara Ann. 2018. *Fractivism: Corporate Bodies and Chemical Bonds*. Durham, NC: Duke University Press.

Wynter, Sylvia. 2003. "Unsettling the Coloniality of Being/Power/Truth/ Freedom: Towards the Human, After Man, Its Overrepresentation—An Argument." *CR: The New Centennial Review* 3 (3): 257–337. https://doi.org /10.1353/ncr.2004.0015.

Zhang, Mengmei, Alfons Buekens, and Xiaodong Li. 2016. "Brominated Flame Retardants and the Formation of Dioxins and Furans in Fires and Combustion." *Journal of Hazardous Materials* 304: 26–39. https://doi.org/10.1016 /j.jhazmat.2015.10.014.

5

ELEMENTAL GHOSTS, HAUNTED CARBON IMAGINARIES, AND LIVING MATTER AT THE EDGE OF LIFE

Astrid Schrader

In the midst of our current environmental crisis and with the naming of the Anthropocene as a new geological epoch that affirms humanity as a geological force, the "microbial turn" in the social sciences has been joined by a "geological turn": a move beyond the biopolitical and "life itself" as analytics of power and existence (Yusoff 2013, 2016).[1] As the Anthropocene provides the context for rethinking the relationships between humanity and the earth, new concepts and theories of *geos*—geologic life (Yusoff 2013, 2016), geosocialities (Palsson and Swanson 2016; Tsing et al. 2017), geopower and geontologies (Povinelli 2016)—emerge in social thoughts, replacing conceptions of life that can be easily opposed to nonlife. "Life" seems to be inadequately accounted for in terms of biological formations alone.

In the recent anthology *Arts of Living on a Damaged Planet: Ghosts and Monsters of the Anthropocene*, Anna Tsing et al. (2017) discuss more-than-human bio- and geosocialities guided by the figures of monsters and ghosts. For them, monsters are entanglements of bodies belonging to different species, chimeras established mainly through symbioses. Simultaneously promising and destructive, "monsters are the wonders of symbiosis and the threats of ecological disruptions" (M2). Ghosts, on the other hand, relate *bios* and *geos*; they "help us read life's enmeshment in landscape" (M2). In addition to entangling life and nonlife, ghosts modify time in the Anthropocene; they "disturb our conventional sense of time, where we measure and manage one thing leading to another" (G9). Unlike monsters, the ghosts that guide through haunted lives and landscapes appear more often than not as traces of past violence rather than as life-enhancing entanglements. For the editors of *Arts of Living on a Damaged Planet,* ghosts announce "a world haunted with the threat of extinction" (G6).

In his contribution to the volume, Nils Bubandt (2017) associates the ghosts of the Anthropocene with a deadly necropolitics; the politics of the current moment is informed by the old ghosts of the carbon-based industry. For Bubandt, the Anthropocene is characterized by a spectrality in that it encourages a retrospective reading of the current impact of humanity (G135). In the Anthropocene, time is out of joint and "life is already geologic" (G136) in a spectral sense, in which the present proceeds from the future, as we are looking ahead toward our own extinction. In spite of this pessimistic outlook, Bubandt ends his discussion of the eruption of a Malaysian mud volcano—caused by humans, through oil drilling—on a hopeful note: there is a promise in the inability "to distinguish the *bios* from the *geos*," which implies "the chance for a novel kind of collaboration between science and the politics of the otherwise" (G137).

I would like to dwell on this promise of the indistinguishability of bios and geos, with the help of scientific accounts of marine viruses and their role in the global carbon circle. In other words, this chapter is an attempt to think of bios and geos together through the transformative agencies of marine viruses, which I call *elemental ghosts.* Moving from "life's enmeshment in landscape" to the entanglement of life and nonlife in the ocean, I am interested in a more hopeful account of ghosts, or a hopeful viral politics. I am after a viral politics, distinct from both a bio- and geopolitics, that is neither catastrophic nor apocalyptic nor that of terrorism, as anthropologist Elizabeth Povinelli proposes (not all viruses are terrorists). What alternatives to terrorism might challenge capitalist extractivisms?

Moving from bios to geos, Povinelli draws our attention to what she calls geontopower that may link bodies to landscapes and topologies. She coins the term *geontology* "to indicate a disruption of a previous formation of power and an analytic placeholder for the formation of power we are living within and through." *Geontology* refers to the interpenetration of biographies (life descriptions) and geographies (nonlife descriptions) and as an analytic for how power is arranged around biogeographical obligation (Povinelli 2014b). While following Povinelli's moves that trouble the opposition between life and nonlife and while paying attention to other possible arrangements of existence that seek to endure, I would like to stay a little longer at the edge of life that is accessible from within the biological or, better, the biogeochemical sciences and explore how an elemental perspective on marine microbes and their viruses may complicate the division between life and nonlife.

THE CARBON IMAGINARY AND GHOSTLY VIRUSES

Povinelli (2016) calls the idea that there is a natural and foundational distinction between life and nonlife our "carbon imaginary." It is her name for a historical move—famously analyzed by Michel Foucault (1970)—that turns a specific biology into ontology.[2] "In the natural, social, and philosophical sciences," Povinelli (2014a) asserts, "'life' acts as a foundational division between entities that have the capacity to be born, grow, reproduce, and die and those that do not. Ontology is, thus, strictly speaking a 'biontology'"—that is, the biological concepts of birth, growth-reproduction, and death are transposed onto the ontological concepts of event, *conatus/affectus*, and finitude of vital materialism (Povinelli 2016, 18). It implies that life is considered the measure for all ethical, social, and political activities and exempts other kinds of entities—such as the transformations of a creek, transgendered in Australian Indigenous dreaming and threatened by mining operations—from ethical considerations with far-reaching social and ecological consequences.

Povinelli's point is that there are forms of existence that cannot just be included in geontological governance but that need to displace the divisions of life and nonlife and challenge late-liberal forms of governance. One example of a mode of existence that Povinelli's Indigenous friends would like to maintain goes by the name of Tjipel. As Povinelli describes it, Tjipel is the transgendered creek, a biogeographical transformation from girl to boy to creek, documenting an encounter of fight and possibly sexual abuse; it names a mode of existence that current Western ontologies cannot account for. In most articulations of Western metaphysics, only living entities have the power to make claims regarding the endurance of their existence. Tjipel does not fit the subjectivity required for the normativity of life; it cannot posit itself; it has no clear beginning or end. Povinelli argues that "Western metaphysics [has become] a measure of all forms of existence by the qualities of one form of existence (*bios, zoe*)" (Povinelli 2016, 5). According to Povinelli, the carbon imaginary maps biological concepts onto ontological ones. I wonder, however, whether these biological concepts that presumably stand in for all forms of life are perhaps specifically human concepts. In other words, the one form of existence by which other forms of existences are measured might not be that of living beings per se but modeled after a specifically human biology and temporality.

From a microbial perspective, the defining processes of bios seem rather anthropocentric; not everyone performs life according to a linear course marked by birth, growth-reproduction, and death. As Lynn Margulis and Dorion Sagan (1995) make clear in their book *What is Life?*, the answer varies with the taxon and scale. For geochemist Vladimir Vernadsky, referring to life as living matter, "life was less a thing and more a happening, a process" (45). "Vernadsky," they write, "portrayed living matter as a geological force—indeed, the greatest of all geological forces. Life moves and transforms matter across oceans and continents. . . . Moreover, life is now known to be largely responsible for the unusual character of Earth's oxygen-rich and carbon dioxide-poor atmosphere" (44). Margulis and Sagan refer here to ancient cyanobacteria that about 2.5 billion years ago began photosynthesizing and created an atmosphere that enabled most contemporary ways of life. For Povinelli (2016, 44), the distinction between life and nonlife also depends on scale and perspective; however, no matter how the division comes about, it is rendered either ontologically significant or irrelevant. Could there be a way to undermine the carbon imaginary from within the sciences?

I am inspired, though, by Povinelli's geontological figure of the Virus. As I trace the movements of elemental carbon in the ocean, viruses serve me as guides, both metaphorically as sociopolitical entities and processes and literally as biogeochemical ones. In other words, I am enrolling viruses here as metaphors, material substances, and biogeochemical processes as I ask what kind of alternative politics may open up if we begin to think with viruses from within the sciences.

The word *virus* has its roots in the Latin and is associated with the venom of a snake, literally meaning "slimy liquid, poison." Just as *bacteria* used to be a synonym for *germs*, before their importance as symbionts in all kinds of organisms was recognized (see Paxson and Helmreich 2014), viruses were until recently studied mainly as disease-causing agents. Their sheer abundance seems to make them essential to our biosphere—there are a hundred million times more viruses on Earth and in the oceans than stars in the universe—but they have long been neglected as important actors in the cycling and recycling of organic matter. New research on marine viruses challenges the view of viruses as mere agents of mortality. Some virologists (e.g., Karen Weynberg) go so far to affirm that the survival of our species depends on them.

Viruses trouble the distinction between life and nonlife; they can be viewed as alternating between living and nonliving phases (Dupré and O'Malley 2009) and between chemical substance and living process, challenging the opposition between substances and processes and organic and inorganic life. They operate on multiple scales (cellular, ecological, and global) simultaneously and are capable of mutating their ontologies. They appear to be in a permanent existential crisis. My point will not be to liberate viruses from their existential crisis; I will rather sharpen and refocus it.

While viruses exist at the edge of life, the central question here is not whether or under what conditions viruses become part of life but rather *how* they trouble the distinction between life (bios) and nonlife (geos). Ghostly marine viruses are not only undoing spatial boundaries between organism and environment, for example, but also temporal boundaries between the alleged futurity of life and the alleged permanent presence of nonlife. In the carbon imaginary, agency is reserved for living beings with specific metabolic processes—birth, growth-reproduction, and death. According to this imaginary, metabolisms create, sustain, and reproduce living beings as some kind of goal-directed intentional substances, Povinelli (2016, 39) argues; the ultimate goal would be to reproduce a version of itself. In other words, life has intentions and potential, while nonlife is pure actuality (44). An alternative imaginary that troubles the boundary between life and nonlife must be able to accommodate new kinds of agencies that do not begin with an intentional unified subject. At stake here is also a refiguring of agency in relation to notions of life, in Isabelle Stengers's (this volume) terms, an exploration of "agency as freed from the opposition between living intentionality and the inorganic."

The turn from bios to geos seems to be following or coinciding with what Paxson and Helmreich (2014) have termed "the microbial turn" that affirms the agency of microbes as part of social relations. Scholarship attributed to either of these turns seeks to expand "the range of agencies permitted to play a part in the construction of social worlds" (Clark and Yusoff 2017, 14). While the geological turn is inspired by the Anthropocene, the microbial turn is inspired by research into the human microbiome, affirming the importance of microbes in and around human bodies for human well-being, while establishing the human as a multispecies ecosystem. In their attempt to account for previously neglected forms of existence and agencies, many articulations within each of the turns remain, in different ways, rather anthropocentric. In articulations of new geosocialities that attend to how "bodies" intertwine with earth systems and deep time histories (Palsson and Swanson 2016, 155), the bodies in question are more often than not human bodies. Heather Swanson's account of biomineralization in salmon bodies seems to be an exception (155). In Nils Bubandt's (2017) example of geosocial entanglements—a human-caused eruption of a Malaysian mud volcano—which demonstrates the indistinguishability between *anthropos* and geos, or the inseparability of geothermal activity from industrial activity (as cause of the volcano's disruption), the bios of biopolitics is most explicitly replaced by the anthropos. Is there a way to think the microbial together with the geological in less anthropocentric terms?

In addition to a remaining or renewed anthropocentrism in articulations of geosocial entanglements, the agencies of earthly materialities are often articulated in terms of liveliness. Palsson and Swanson (2016) speak paradoxically of the liveliness of nonlife (the stone or the mineral). In this context, Nigel Clark asks why we continue to highlight the liveliness of the inorganic at the expense of other properties of minerals (Yusoff 2013, 789). Kathryn Yusoff (2013, 790) associates a "vital biopolitics" with an "organic chauvinism." I argue that viruses, as living matter at the edge of life, modify the link between liveliness and material agency. Rather than conflating agencies with liveliness and including ever more kinds of agencies into social bodies, I am more interested in how specific readings of marine science may contribute to the opening up of new ontologies and political spaces through the reconfiguration of the normativity of the living.

In drawing attention to how geology has become political, Yusoff (2013, 784) notes that the Anthropocene "creates a new geologic subject, defined by its use of fossil fuels." In other words, there is "a geologic dimension of subjectivity" that acknowledges the activity of fossil fuels within contemporary corporeality (789). Yusoff's point is the disavowal of the agency of fossil fuels in the formation of subjectivity in the Anthropocene. Fossil fuels are, however, not the only possible geosocial relation in which carbon participates: elemental carbon does not have to turn into the fuel of capitalist modes of production, viewing all things as profitable and vital, necessarily relying on the distinction between life and nonlife. Elementary carbon can be quite recalcitrant or refractory.

In the Anthropocene, the apparent self-evident distinction between life and nonlife, which Povinelli (2016) calls our carbon imaginary, begins to crumble. The necessity to consider a multitude of temporalities, including deep time, is shaking the old carbon imaginary. While not exactly replacing the old ghosts of the carbon-based industries, but perhaps accompanying them, new ghostly agencies manifest themselves in the elemental recycling of carbon in the ocean. Reading elemental theory together with scientific accounts of ecologies of marine viruses and their role in the global carbon cycle, this chapter seeks to develop an alternative carbon imaginary that does not figure carbon as the essence of life and capitalist production but traces it through elemental cycling with the help of viruses, in which ontologies are allowed to mutate between elemental substance, process, and relation.

VIRAL POLITICS AND THE FIGURES OF GEONTOPOWER

According to Povinelli, the carbon imaginary can no longer adequately account for power formations in an age of climate change. The anthropocentric figures that emerged from a Foucauldian biopolitics need replacement with figures that more accurately present contemporary formations of power. Povinelli (2016, 15) proposes to replace the all-too-human sexual figuration (the hysterical woman, the Malthusian couple, the perverse adult, and the masturbating child) with the Desert, the Animist, and the Virus. These figures are considered symptoms and diagnostic tools of late-liberal governance; they emerge from contemporary formations of power. One can think of these figures as "a collection of governing ghosts" existing in between two worlds, one in which the distinction between life and nonlife is dramatized and another in which it is no longer relevant (16). I argue that Povinelli's collection of ghosts does not only exist between the living and nonliving but can also be read as intervening in a clear distinction between the sociopolitical and the technoscientific world.

The Desert is most clearly associated with capitalist modes of production. The Desert is denuded of life but can, with proper technology, be made hospitable again. Life in the Desert is always in danger of turning into nonlife. "The Carbon Imaginary lies at the heart of this figure" (Povinelli 2016, 16); it implies terror of the other of life. While the Desert usually refers to an ecosystem that lacks water, it becomes, for Povinelli, an affect that motivates the search for life. The difference between life and nonlife is essential for that figure. The Desert also operates within fossils that have lost life but can, upon extraction as a form of fuel, provide the conditions for specific forms of life in contemporary capitalism, accompanied by new forms of mass deaths and extinctions.

"At the heart of the figure of the Animist lies the imaginary of the Indigene" (Povinelli 2016, 17). For the Animist, all forms of existence are alive, vitally animated, and affective. The Animist is associated with a philosophical vitalism expressed in Spinoza's principle of *conatus* (the striving to preserve in being) and *affectus* (the ability to affect and be affected). The Animist collapses life and nonlife; the figure is employed to delegate Indigenous groups to premodern status, where they see life or agency, subjectivity and intentionality, while others see only nonlife.

Finally, the Virus is associated with the Terrorist, an active antagonist, which Povinelli associates with the Ebola virus, the waste dump, and zombies, disrupting the distinction between life and nonlife; after 9/11 it becomes associated with fundamentalist Islam and the radical Green movement (Povinelli 2016, 19) but also with her own project with the Karrabing Film Collective, an Indigenous corporation in Australia, which explicitly rejects state forms of land tenure and group recognition. The noncompliant Virus dwells in an existential crisis (19). While the politics of the Desert is associated with the contemporary mode of capitalist production, accompanied by extinction events, the Animist accounts for Aboriginal land claims; viral politics is associated with terrorism.

Povinelli's description of the figure of the Virus is clearly informed by the biology of viruses, whose status as a "form of life" has never been settled by philosophers of biology. As viruses do not reproduce or metabolize by themselves—they hijack the functionality of their hosts to perform these life-sustaining functions—they do not qualify as living beings, since "life" is assumed to be self-organized and self-contained, even though many biologists speak of the life cycle of a virus and some refer to them as nonliving life forms (see, for example, Brussaard 2019).

Povinelli writes that the Virus "can use and ignore this division for the sole purpose of diverting the energies of arrangements of existence in order to extend itself. The Virus copies, duplicates, and lies dormant even as it continually adjusts to, experiments with, and tests its circumstances. It confuses and levels the difference between Life and Nonlife while carefully taking advantage of the minutest aspects of their differentiation" (Povinelli 2016, 18–19). Linking the Virus to the Terrorist, she asserts, "The Virus may seem to be the radical exit from geontopower at first glance, to be the Virus is to be subject to intense abjection and attacks, and to live in the vicinity of the Virus is to dwell in an existential crisis" (19). The biology that allows the association of viral behavior with terrorism needs an update, however. Viruses infect not only human beings but all living organisms, including, most importantly, bacteria. In the ocean, bacteriophages—viruses specialized to infect specific bacteria—kill so many microbes that their remains divert a significant amount of carbon from the food chain.

Through the infection of bacteria, marine viruses play a major role in regulating and manipulating the carbon cycle in the ocean. Paying close attention to the biology and biogeochemistry of marine viruses may offer an alternative exit from geontopower, the possibility of a viral politics other than terrorism. What may happen if we combine Povinelli's predominantly sociopolitical figure with a biogeochemical one? In what other ways could viruses intervene in geontopower? Rather than residing between two worlds—one in which the distinction between life and nonlife matters and another in which that distinction is irrelevant—as ghostly figures *and* actual processes, marine viruses reconfigure that very division and redistribute agencies across the divide. Diffracting Povinelli's figure of the Virus with elemental theory may transform the carbon imaginary into a haunted one.[3]

ELEMENTAL RELATIONS

The pre-Socratic philosopher Empedocles (490–430 BCE) is held responsible for the cosmogenic theory of the four classical material elements: earth, air, fire, and water (Aristotle added a fifth one, ether). The four elements are moved by two opposing affective forces, love and strife, which are attractive and repulsive, respectively. Love (φιλότης) is responsible for the attraction of different forms of matter, and strife (νεῖκος) is the cause of their separation. If these elements make up the universe, then love and strife explain their variation. In Empedocles's cosmic cycle, life emerges only when the two affective forces are in tension or in relative movement to each other. Purity in either is associated with the absence of life.

Elemental theory has recently been taken up and brought into the environmental humanities by Jeffrey Cohen and Lowell Duckert (2015), with the help of an anthology titled *Elemental Ecocriticism*, which is committed to "an ecomaterialism that conjoins thinking the limits of the human with thinking elemental activity and environmental justice." According to Cohen and Duckert, elemental theory retheorizes matter as a precarious system and a dynamic entity rather than as a resource or commodity. Elemental theory should not be seen as nostalgia for the Empedoclean elements but rather as a provocation for environmentality as a "disanthropocentric reenvisioning of the complicated biomes and cosmopolities within which we dwell" (5). For Cohen, the classical elements are partners in world making; they are finite (even mortal); they push against borders and spark couplings. They are the most promising compositions of the world (see also Helmreich, this volume).

Elemental theory attempts to address the complexity of life beyond the organic. For Cohen (2011), following the four classical elements—earth, water, air, and fire—"in their movements suggests a method of thinking about the world that is atheological and unpredetermined." He calls it a secular and speculative vitalism. However, as the contributions to *Elemental Ecocriticism* demonstrate, in such a following or tracing of the elements, the division between vitalism and materialism is anything but clear. Ancient elemental theory offers a mode of understanding materiality that does not center the cosmos around the human (Cohen and Duckert 2015); rather than revealing the vitality of inert substances, an elemental thinking begins from within the entanglements of life and nonlife and challenges the opposition between substances and processes.

Elemental relations are characterized by a paradoxical entanglement between separation and combination (see Mentz 2015), which Karen Barad (2014) calls a "cutting together-apart." They are both discrete and conjoined. For Cary Wolfe (2015, 290), they exist at the edge of relations: "In their striving to be discrete they only trace all the more the arc by which we know them to be elements." Elemental thinking offers more than the reanimation of matter; it can hold a multiplicity of kinds of agencies together. Elemental relations provide potential for the impossible, the imaginary, and the abandoned, like viruses that critically transform the elemental associations of carbon, the stuff of life. Also like viruses, elements hold metaphors and materiality in complementary tension. Cohen describes them as "metaphor magnets": they yearn to be metaphors; they are *matterphors* (Cohen and Duckert 2015, 11).

Rather than simply offering a new way to reanimate matter, elemental thinking offers a possibility to think together substance and process, sensuous objects and material relations, the elemental as a physiochemical entity and as "a state of matter from which every other material is derived" (McCormack 2017, 421). It is also conceived of as "the question of where speculative realities begin: with process, with object, with thing, with event or with movement" (421). I am intrigued by the multiplicity and indeterminacy of elemental ontologies and ask how an elemental thinking, with the help of viruses as matterphors (substance and word combined), can develop an alternative and less anthropocentric carbon imaginary.

According to the philosopher John Sallis (2012), the elemental turn signifies a return to "nature," but a nature in which "natural things" can be distinguished from the "elements of nature." Elements such as earth and air are different from individually delimited things.[4] "We" (humans) relate differently to them: while "we comport ourselves to things across a certain distinct interval . . . we are encompassed by elements" (348). Elements are not sensible things but *of* the sensible; air, for example, is not visible but is what makes visibility possible. Similarly, viruses are *of* the sensible; they are "natural elements"—both substance and process, agents of mortality, diversity, and transformations of spaces and times. They are elemental ghosts.

Bernadette Bensaude-Vincent describes the history of chemistry in terms of an ongoing confrontation between two doctrines: atomism and elemental theory in the ancient classical world (Bensaude-Vincent and Simon 2012). Among many other differences, matter is discrete in one case and continuous in the other; viruses are both. Alive or not, viruses are like elements: "animated materialities with and through which life thrives" (Cohen and Duckert 2015, 13). Carbon, argue Sacha Loeve and Bensaude-Vincent (2017, 190), "connects modern chemistry with ancient metaphysics." Carbon served Dmitri Mendeleev, the inventor of the periodic table, as a template to define chemical elements. It was considered a hypothetical, abstract, basic substance, something that would not alter in chemical transformations. Unlike simple substances, such as diamond, graphite, and coal—the concrete (allotropic) forms of carbon—a basic substance was considered to have no existence in space and time. "While simple substances come into existence as concrete and physical entities at the end of a process of analysis and purification, elements are the material but invisible parts of simple and compound bodies. Carbon is a hypothetical abstract entity since it can never be isolated. . . . Nevertheless the element carbon is real and identifiable by a positive individual feature: its atomic weight . . . 12" (190). Carbon as substance was simultaneously a concrete phenomenal entity and a universal substrate capable of explaining changing phenomena. Elemental carbon combines the idea of the discrete atom with that of a continuous and enduring (classical) element, something that circulates without origin.

Tracing carbon through a number of incarnations or signatures, Loeve and Bensaude-Vincent argue that carbon calls for multiple systems of knowledge, invites ontological pluralism, and encourages modes of existences to interact. Carbon is, among many other things, memory —ancient buried sunshine, the material memory of life on Earth. "The layers of biomass accumulated over millions of years as sediments of hydrocarbons are like a repository of the past"—memories that have been transformed by *homo economicus* "at a rate of a few centuries per year" (Loeve and Bensaude-Vincent 2017, 194). It may require the agency of some elemental ghosts to transform the temporalities of these material memories.

VIRUSES AS "BEINGS" AT THE EDGE OF LIFE

Being made out of carbon is, of course, not a sufficient criterion for life. Biology textbooks list up to seven criteria, the most important being the capacities to metabolize and to reproduce. Just like that of carbon, the liveliness of viruses depends on their connection.

Considered as a substance, a virus is basically a piece of DNA or RNA wrapped in proteins (which they do not produce themselves); until it finds a living host to infect, it is merely a piece of chemical information. Viruses are not generally considered organisms; they do not consume or metabolize. They do not reproduce by themselves, as they need a cellular host to multiply. They are not living things. They are not non-living things completely either. Viruses seem to have no problem alternating between living and nonliving phases, depending on circumstances. Viruses are said to come alive when they invade a living cell of the host—that is, at least as long as life is defined in terms of reproduction and requires an autonomous unit or individual that produces and reproduces itself. But "the idea of an organism as the basic unit of life is likely to have had its day" (McRae n.d.). The virus that can exist in different phases or stages is sometimes confused with the virion—the infective particle phase of a virus outside a host cell, which is only a particular life stage of the "virus." "Viral genes are no more living or dead than any other genes" (McRae n.d.).

Why should autonomy and individuality be essential criteria for life? In light of the ubiquity and importance of symbiosis for both animal development and evolutionary transitions, autonomy, individuality, and self-maintenance have become questioned as criteria for the living (Gilbert 2014). Philosophers of biology John Dupré and Maureen O'Malley (2009) consider cooperation a central characteristic of living matter. They do not restrict aliveness to cells or organisms; rather, they see life emerging when species (or lineage-forming entities) collaborate in metabolism, forming functional wholes. From this perspective, viruses can be viewed as alternating between living and nonliving phases, between chemical substance and living process (see also Dupré and Guttinger 2016). If multispecies cooperation and codependencies are the norm rather than the exception for the maintenance of life, then there is no reason not to regard viruses as part of life.

In an essay titled "The Ecological Virus," O'Malley (2016) takes a slightly different view. Rather than emphasizing cooperation, she highlights the distributed ecological agency of viruses over a possible biological agency. Neither ecological nor biogeochemical actors have to be organisms. The ecological virus is not a tiny organism that aspires to cooperation, nor does it participate in a division of labor for the good of the community, asserts O'Malley. Rather, viruses appear to create the conditions to replace themselves, while they are nonorganismal, nonmetabolic, and noncooperative. In her virocentric ecology, viruses are becoming distributed agents associated with functional types. Such a view shifts our attention not from bios to geos but from the actions of organisms, individual actions, and concentrated subjectivity to distributed agencies that link viruses to the global climate system. From such a perspective, viruses become utterly indifferent to the distinction between life and nonlife. What matters is not so much what they are but what they do. Life is understood in terms of interactions of multiple biological and biogeochemical agents, "not all of which are living things" (78).

VIRUSES AS MAJOR PLAYERS IN THE GLOBAL CARBON CYCLING

Viruses are not only metaphors for the "in-between"—between life and nonlife—or for the irrelevance of that distinction; marine viruses are also major players in the carbon production and recycling, in the transforming of living matter and temporalities. Through the infection of bacteria, marine viruses not only alter the rate of carbon fixation—the conversion of atmospheric CO_2 into bodily energy by photosynthesizing bacteria—but also play a major role in regulating and manipulating the conversion between organic and inorganic carbon in the ocean. In addition, they enable and contribute to the storage of carbon in the ocean.

For those scientists who still conceive of the earth systems in terms of giant clockworks, bacteria are the "cogs" and viruses are the "lubricants" of that machine (Willie Wilson, director of the Marine Biological Association, personal communication with the author, February 4, 2015). They act as vectors of genetic transfers, driving diversity in the ocean. They shape populations, drive evolution, and even influence the weather; according to virologist Karen Weynberg (2016), without viruses life on Earth would come to a halt.

As the main driver of mortality in the ocean and recycler of organic matter, they exert great influence on community structures. In releasing nutrients from their hosts, marine viruses contribute to all of the elementary nutrient cycles. Their interactions with bacteria have major implications for marine food webs and the biogeochemical cycles; the interplay between viruses and microbes is crucial "for an understanding of how biotic and abiotic factors maintain living systems" (O'Malley 2016, 71). Ironically, though, the very destruction wrought by viruses contributes to the maintenance of the world as microbial; as viruses more often than not go for the fastest reproducers, they are also agents of diversity. The paradox is resolved if viruses are considered to be distributed ecological agents rather than biological agents. While viruses usually kill their hosts, they can be beneficial to the population.

There is about fifty times more carbon in the ocean than in the atmosphere (Rackley 2010). The ocean acts as a carbon sink for anthropogenic atmospheric carbon dioxide (CO_2), thus mitigating global warming.[5] Carbon enters the ocean in one of two ways: via direct absorption from the atmosphere as carbon dioxide, which then dissolves and reacts with seawater to form carbonic acid and carbonate—a process that reduces the PH level of the seawater and leads to ocean acidification—or via photosynthesis, whereby microbes convert CO_2 into organic carbon and oxygen. "Most carbon enters the biological pool via photosynthesis" (Wilhelm and Suttle 1999, 783).

The biological mechanism for long-term carbon storage in the ocean is called the *biological pump*, a term that describes the flux of particulate organic matter to the ocean floor. However, only 0.1 percent of the organic carbon produced on the surface of the ocean makes it to the bottom; the rest is eventually respired back to CO_2. Within a few decades, most of the organic carbon will be returned to dissolved inorganic carbon (Jiao et al. 2010). Scientists now believe that another mechanism—the microbial carbon pump (MCP)—sequesters carbon much closer to the ocean surface "in the form of long-lived dissolved organic matter, which is resistant to biological decomposition and assimilation" (Wang 2018, 287).

Considering that 98 percent of living matter in the ocean consists of microbes, it is astounding how long their activities have been neglected in models of global carbon cycles. Photosynthesizing bacteria are not only the so-called primary producers of organic carbon in the ocean—and indeed, they produce half of the oxygen on the planet—they are also major recyclers of organic matter. It was previously thought that bacteria in the ocean were exclusively regulated by grazers—that is, by their predators in the so-called food chain, mainly larger microbes such as zooplankton. In this way, the photosynthetically produced carbon would simply move up the food chain to higher trophic levels. Now it is estimated that about 40 percent of them die due to viral infections.

This linear model of carbon transport had to be corrected twice, as the roles of heterotrophic bacteria (those that eat other critters made out of carbon) and of viruses were increasingly recognized. First, the so-called microbial loop was discovered (Fenchel 2008), which was further modified by the action of viruses called the viral shunt. When viruses kill bacteria, they contribute to a vast store of dissolved organic carbon

(DOC) that cannot be easily taken up by larger organisms but can be consumed by heterotrophic bacteria. This DOC includes zooplanktons' liquid wastes and a jellylike substance ("cytoplasm") that leaks out of phytoplankton cells. When these many bacteria are later eaten by micrograzers such as flagellates and ciliates, the formerly "lost" carbon and energy are recycled back into the marine food web. In this way, viruses divert the upward flow of carbon and nutrients from secondary consumers, increasing the recycling of organic material. Wilhelm and Suttle (1999, 785) estimate that "as much as one-quarter of the organic carbon flows through the viral shunt." It is not entirely clear how exactly the increased recycling of carbon in the upper ocean affects the amount of CO_2 in the air and the possibility of carbon transport into the deep ocean. In diverting the carbon flow, viruses may alter the efficiency of the biological pump. There are, however, arguments for either direction of that alteration (Breitbart et al. 2018). Currently marine scientists are more interested in the other carbon sequestration mechanism, the MCP, to which viral lysis contributes significantly. The MCP manipulates the lifetime of organic carbon in the ocean: it "contributes to carbon sequestration through the biochemical transfer of carbon from organic compounds with a lifetime <100 years to DOC fractions with a lifetime >100 years" (Legendre et al. 2015, 442).

The microbial loop is responsible for the recycling of carbon in relatively short terms close to the ocean surface. Heterotrophic microbes remove dissolved organic carbon (DOC) from the water and make more microbes, which then become food for larger organisms. In this way, the microbes convert DOC into particulate organic carbon (POC)—the bacteria themselves—thereby returning organic carbon back to the food web as particulate organic matter.

Dissolved organic carbon should not be confused with dissolved inorganic carbon, which contributes to ocean acidification. In fact, DOC is not really dissolved at all but contains very small particles that pass through a filter with a pore size of 0.2 μm. So, the difference between DOC and POC, the material that does not pass through the filter, is defined operationally. Even though the difference is defined rather arbitrarily, DOC and POC follow rather different pathways through the carbon cycle. While much of the POC is transferred directly to high trophic levels, DOC can be taken up only by microbes and is recycled in the microbial loop (Wilhelm and Suttle 1999). Therefore, the microbial loop removes parts of the DOC from the water and converts it to POC for consumption by bigger critters. Some DOC, however, is left behind; scientists call it recalcitrant or refractory DOC (RDOC), which accumulates over time. The microbial carbon pump is seen as a framework of mechanisms that remove "DOC from the short-term carbon flow into another pool, the refractory DOC, which keeps the carbon in the water column for a long time" (Wang 2018, 290). Thus, the MCP describes a temporal process, converting short-term carbon flow into long-term carbon storage. Refractory DOC can store carbon for six thousand years; it accumulates quite slowly, and this large pool of carbon is relatively inert. This inert, inedible carbon constitutes another way of carbon sequestration in the ocean. It is still a mystery why a huge and diverse pool of microbes cannot make use of the refractory DOC, but it is definitely fortunate: if they could take it up and all RDOC were respired, CO_2 in the atmosphere would double (Azam and Jiao 2011). The MCP has a potentially huge impact on climate change, but scientists are still hesitant with any predictions (Jiao and Zheng 2011).

While most marine scientists agree that "virus infections of microbes could change the flux of carbon and nutrients on a global scale" (Toon 2016), the "predictions about the interactions of viruses, ecosystems, geochemical cycles, and climate change are still vague" (O'Malley 2016, 76). Viruses influence the global carbon cycle, but to what effect they do so remains uncertain. Moreover, the action of viruses seems to be surprisingly local—that is, dependent on scale and context: "Researchers have found that viral populations vary dramatically from location to location, and at differing depths in the sea. The study highlights another source of uncertainty governing climate models and other biogeochemical measures" (Wigington et al. 2016).

CONCLUDING WITH MONSTERS, GHOSTS, AND A HAUNTED CARBON IMAGINARY

The "cooperative virus" of Dupré and O'Malley's (2009) account seems to be a monster in Tsing et al.'s (2017) sense of the monstrous, thriving in the symbiopolitical realm. The politics associated with such a cooperative monster could be called a symbiopolitics. Coined by Stefan Helmreich, symbiopolitics emphasizes that the governance of life no longer solely proceeds through the management of individual bodies or population but through the governance of relations among entangled living things (Helmreich 2009, 15). However, this viral monster seems to be too easily domesticated as the conception of life is expanded; it does not seem to alter the relationship between life and nonlife and therefore does not threaten the established order of geontological governance.

Marine viruses exist only on the move; their ontology depends on the scale and the specific questions that are put to them; they can be destructive substances, a living process, or a distributed and transformative agent. Maureen O'Malley's (2016) notion of distributed agency in a virocentric ecology modifies the stubborn link between liveliness and agency. Acting as agents of mortality on one scale, viruses contribute to an increase of diversity on another scale.

My discussion of marine viruses as transformative agents in the carbon cycle suggests that carbon cycles in the ocean do not easily lend themselves to a "carbon imaginary" that supports the endless activities of capitalist production. Like the elemental breakdown in the soil, viruses "delinearize the cosmological tale," in the words of María Puig de la Bellacasa (this volume); they "turn origins into cycles, lines into spirals, but also . . . [and] complicate neat cyclic models, with twists and overlaps." Unlike the bacteria in the soil in Puig de la Bellacasa's account, however, viruses do not always prepare elements for reuse; instead, they produce a huge amount of "'inedible' organic carbon" (Stone 2010, 1476). This breakdown—this useless, elemental waste—seems critical for human survival.

While marine viruses are clearly zombies (living dead), more importantly, perhaps, they are also ghosts, specters of inheritance and transformations. They are not only impartial to the distinction between life and nonlife; they also transform the temporalities of earthly matter and memories. Transforming the flux of short-lived carbon into long-lived unproductive carbon waste, they engender refractory dissolved carbon as ghosts of former organic production. Future potential (of life) and pure actualities (of nonlife) are becoming blurred when "inedible" unproductive carbon appears to be buying humanity some "time," apparently postponing our imminent extinction.

In his book *Specters of Marx*, Jacques Derrida (1994) formulates a "logic of haunting" that transcends "the opposition between presence and non-presence, actuality and inactuality, life and non-life" (12). "To haunt does not mean to be present," he writes, "and it is necessary to introduce haunting into the very construction of a concept. Of every concept, beginning with the concepts of being and time" (161). Ghosts undo the metaphysics of presence; they are figures of justice that are essential for transformations of and in "time." As Derrida asserts, "Present existence or essence has never been the condition, object, or the thing [chosen] of justice" (175). I find hope in the alterity and ghostly undecidability of what viruses are, their ontologically indeterminacy, and the simultaneous demand to account for their existence and actions in order to get a better picture of global warming and ocean acidification.

Just because industrial capitalism depends on the transformation of "ancient sunshine" (Povinelli 2016, 10), humans do not have to (and cannot) continue "consuming that memory at a rate of a few centuries per year" (Loeve and Bensaude-Vincent 2017, 194). The old ghosts of the carbon industries do not determine all possible futures. As elemental ghosts, viruses, and Derrida (1994) teach us, inheritance is always a task. While ghostly marine viruses exist only on the move, like Empedocles's root elements, they may also contribute to balance and stability in a world of perpetual flux. "Supposedly outdated articulations of elemental activity . . . can propel care, grasp, and justice" (Cohen and Duckert 2015, 4).

If the old carbon imaginary asks questions about beginnings and ends, birth and reproduction, apocalypse and rebirth, a haunted, transformed one may start in the middle; like Povinelli's various forms of geolife that seek to endure, we may ask questions about orientations, directions, and connections. A haunted carbon imaginary acknowledges that divisions between life and nonlife sometimes matter, but ontologies may mutate with the help of elemental ghosts that distribute agencies across a divide that is no longer fundamental. With Yusoff (2016), I would like to affirm that thinking with viruses as elemental ghosts, as agents of elemental carbon transformation, does not terminate in the Anthropocene; it sets off toward different, and perhaps more hopeful, indeterminate futures.

NOTES

1. While the Anthropocene was originally introduced as a scientific (geological) term by the microbiologist Eugene Stoermer in the 1980s and popularized by the Nobel Prize–winning chemist Paul Crutzen, it took on a life of its own in the humanities and social sciences as a shorthand for discourses on the current environmental crisis. Technically, the Anthropocene refers to the current geological age, viewed as the period during which human activity has been the dominant influence on climate and the environment (Crutzen 2002).

2. Foucault (1970) claimed that "life itself did not exist" (128) before the end of

the eighteenth century. Life was brought into existence with the emergence of biology and relied on the discovery of the organic structure of living things: "life appeared as the effect of a patterning process—a mere classifying boundary" (268). At the same time, when organic structure became associated with specific functions and became the guiding principle for classification, the opposition of the organic and inorganic and of life and nonlife became fundamental. Foucault writes: "The organic becomes the living and the living is that which produces, grows, and reproduces; the inorganic is the non-living, that which neither develops nor reproduces; it lies at the frontiers of life, the inert, the unfruitful—death" (232).

3. After Barad (2007), *diffraction* denotes a methodology, a reading practice, in which theories are read through each other (rather than against each other) and made to interfere productively.

4. Sallis (2012) confusingly refers to "sky" rather than "air."

5. As a global average, the oceans take up about 2 percent more of the gas than they release.

REFERENCES

Azam, Farooq, and Nianzhi Jiao. 2011. "Revisting the Ocean's Carbon Cycle." In *The Microbial Carbon Pump*, edited by Nianzhi Jiao, Farooq Azam, and Sean Sanders. Science AAAS Business Office. https://www.sciencemag.org/site /products/scor_aaas.pdf.

Barad, Karen. 2007. *Meeting the Universe Halfway: Quantum Physics and the Entanglement of Matter and Meaning*. Durham, NC: Duke University Press.

Barad, Karen. 2014. "Diffracting Diffraction: Cutting Together-Apart." *Parallax* 20 (3): 168–87.

Bensaude-Vincent, Bernadette, and Jonathan Simon. 2012. *Chemistry: The Impure Science*. London: Imperial College Press.

Breitbart, Mya, Chelsea Bonnain, Kema Malki, and Natalie A. Sawaya. 2018. "Phage Puppet Masters of the Marine Microbial Realm." *Nature Microbiology* 3 (7): 754–66.

Brussaard, Corina. 2019. "The Ecological Importance of Marine Viruses." Lecture sponsored by the Royal Swedish Academy of Sciences, May 23, 2019. https:// www.youtube.com/watch?v=2RFtYd_gQnU.

Bubandt, Nils. 2017. "Haunted Geologies: Spirits, Stones, and the Necropolitics of the Anthropocene." In *Arts of Living on a Damaged Planet: Ghosts and Monsters of the Anthropocene*, edited by Anna Tsing, Heather Swanson, Elaine Gan, and Nils Bubandt, G121–41. Minneapolis: University of Minnesota Press.

Clark, Nigel, and Kathryn Yusoff. 2017. "Geosocial Formations and the Anthropocene." *Theory, Culture and Society* 34 (2–3): 3–23.

Cohen, Jeffrey Jerome. 2011. "An Abecedarium for the Elements." *Postmedieval* 2:291–303.

Cohen, Jeffrey Jerome, and Lowell Duckert, eds. 2015. *Elemental Ecocriticism: Thinking with Earth, Air, Water, and Fire*. Minneapolis: University of Minnesota Press.

Crutzen, Paul J. 2002. "Geology of Mankind." *Nature* 415 (6867): 23.

Derrida, Jacques. 1994. *Specters of Marx: The State of the Debt, the Work of Mourning, and the New International*. Translated by Peggy Kamuf. New York: Routledge.

Dupré, John, and Stephan Guttinger. 2016. "Viruses as Living Processes." *Studies in History and Philosophy of Science Part C: Studies in History and Philosophy of Biological and Biomedical Sciences* 59:109–16.

Dupré, John, and Maureen A. O'Malley. 2009. "Varieties of Living Things: Life at the Intersection of Lineage and Metabolism." *Philosophy and Theory in Biology* 1.

Fenchel, Tom. 2008. "The Microbial Loop—25 Years Later." *Journal of Experimental Marine Biology and Ecology* 366 (1): 99–103.

Foucault, Michel. 1970. *The Order of Things: An Archaeology of the Human Sciences*. New York: Pantheon.

Gilbert, Scott F. 2014. "Symbiosis as the Way of Eukaryotic Life: The Dependent Co-Origination of the Body." *Journal of Biosciences* 39 (2): 201–9.

Helmreich, Stefan. 2009. *Alien Ocean: Anthropological Voyages in Microbial Seas*. Berkeley: University of California Press.

Jiao, Nianzhi, Gerhard J. Herndl, Dennis A. Hansell, Ronald Benner, Gerhard Kattner, Steven W. Wilhelm, David L. Kirchman, et al. 2010. "Microbial Production of Recalcitrant Dissolved Organic Matter: Long-Term Carbon Storage in the Global Ocean." *Nature Reviews Microbiology* 8 (8): 593–99.

Jiao, Nianzhi, and Qiang Zheng. 2011. "The Microbial Carbon Pump: From Genes to Ecosystems." *Applied and Environmental Microbiology* 77 (21): 7439–44.

Legendre, Louis, Richard B. Rivkin, Markus G. Weinbauer, Lionel Guidi, and Julia Uitz. 2015. "The Microbial Carbon Pump Concept: Potential Biogeochemical Significance in the Globally Changing Ocean." *Progress in Oceanography* 134:432–50.

Loeve, Sacha, and Bernadette Bensaude-Vincent. 2017. "The Multiple Signatures of Carbon." In *Research Objects in Their Technological Setting*, edited by Bernadette Bensaude-Vincent, Sacha Loeve, Alfred Nordmann, and Astrid Schwarz, 185–200. London: Routledge.

Margulis, Lynn, and Dorion Sagan. 1995. *What Is Life?* New York: Simon and Schuster.

McCormack, Derek P. 2017. "Elemental Infrastructures for Atmospheric Media: On Stratospheric Variations, Value and the Commons." *Environment and Planning D: Society and Space* 35 (3): 418–37.

McRae, Mike. n.d. "Are Viruses Alive?" ScienceAlert. Accessed March 4, 2018. https://www.sciencealert.com/are-viruses-alive.

Mentz, Steve. 2015. "Phlogiston." In *Elemental Ecocriticism: Thinking with Earth,*

Air, Water, and Fire, edited by Jeffrey Jerome Cohen and Lowell Duckert, 55–76. Minneapolis: University of Minnesota Press.

O'Malley, Maureen A. 2016. "The Ecological Virus." *Studies in History and Philosophy of Science Part C: Studies in History and Philosophy of Biological and Biomedical Sciences* 59:71–79.

Palsson, Gisli, and Heather Anne Swanson. 2016. "Down to Earth Geosocialities and Geopolitics." *Environmental Humanities* 8 (2): 149–71.

Paxson, Heather, and Stefan Helmreich. 2014. "The Perils and Promises of Microbial Abundance: Novel Natures and Model Ecosystems, from Artisanal Cheese to Alien Seas." *Social Studies of Science* 44 (2): 165–93.

Povinelli, Elizabeth A. 2014a. "Geontologies of the Otherwise." Theorizing the Contemporary, Fieldsights, January 13. https://culanth.org/fieldsights/geontologies-of-the-otherwise.

Povinelli, Elizabeth A. 2014b. "On Biopolitics and the Anthropocene: Elizabeth Povinelli, interviewed by Kathryn Yusoff and Mat Coleman." *Society and Space*, March 7.

Povinelli, Elizabeth A. 2016. *Geontologies: A Requiem to Late Liberalism*. Durham, NC: Duke University Press.

Rackley, Stephen A. 2010. "Ocean Storage." In *Carbon Capture and Storage*, 267–86. Boston: Butterworth-Heinemann.

Sallis, John. 2012. "The Elemental Turn." *The Southern Journal of Philosophy* 50 (2): 345–50.

Stone, Richard. 2010. "The Invisible Hand Behind A Vast Carbon Reservoir." *Science* 328 (5985): 1476–77.

Toon, John. 2016. "Study Shows Large Variability in Abundance of Viruses That Infect Ocean Microorganisms." Georgia Tech News Center, January 25. https://www.news.gatech.edu/2016/01/24/study-shows-large-variability-abundance-viruses-infect-ocean-microorganisms.

Tsing, Anna, Heather Swanson, Elaine Gan, and Nils Bubandt, eds. 2017. *Arts of Living on a Damaged Planet: Ghosts and Monsters of the Anthropocene*. Minneapolis: University of Minnesota Press.

Wang, Ling. 2018. "Microbial Control of the Carbon Cycle in the Ocean." *National Science Review* 5 (2): 287–91.

Weynberg, Karen. 2016. "What You Didn't Know about Viruses in the Ocean." Filmed November 3, in Townsville, Australia. TEDx video, 14:04. https://www.youtube.com/watch?v=DDyt3jbCPPg.

Wigington, Charles H., Derek Sonderegger, Corina P. D. Brussaard, Alison Buchan, Jan F. Finke, Jed A. Fuhrman, Jay T. Lennon, et al. 2016. "Re-Examination of the Relationship between Marine Virus and Microbial Cell Abundances." *Nature Microbiology* 1 (3): 15024.

Wilhelm, Steven W., and Curtis A. Suttle. 1999. "Viruses and Nutrient Cycles in the Sea: Viruses Play Critical Roles in the Structure and Function of Aquatic Food Webs." *BioScience* 49 (10): 781–88.

Wolfe, Cary. 2015. "Elemental Relations at the Edge." In *Elemental Ecocriticism: Thinking with Earth, Air, Water, and Fire*, edited by Jeffrey Jerome Cohen and Lowell Duckert, 286–97. Minneapolis: University of Minnesota Press.

Yusoff, Kathryn. 2013. "Geologic Life: Prehistory, Climate, Futures in the Anthropocene." *Environment and Planning D: Society and Space* 31 (5): 779–95.

Yusoff, Kathryn. 2016. "Anthropogenesis: Origins and Endings in the Anthropocene." *Theory, Culture and Society* 33 (2): 3–28.

6   THE ARTIFICIAL WORLD

Joseph Masco

What is a world actually made of? As a start, we might say that love, pain, and desire are fundamentals, followed quickly by fantasy, imagination, and hope. In this strange historical moment of late-industrial neoliberalism, many believe that the market makes reality, sorting and optimizing, choosing winners and losers with a vast, invisible hand. Many more would say the world is made out of the everyday struggle to overcome and survive the historical and amplifying force of capitalism itself—of extraction machines, bought and sold bodies, ongoing dispossessions, anti-Blackness, and investments in perpetual war (see Liboiron 2021; Masco 2020; Sharpe 2016; Stengers 2015). Any way of approaching this question of worldly essence is therefore superbly complex, filled with process and artifact, transformation and sacrifice, amnesia and unknowns. But a world is, of course, always more than its base materiality: it is sensory, affectively charged, and mutable (see Leslie 2005). A world involves a way of living as much as the material fact of existence, involving the qualities, investments, and blindness of a people, an economy, an era. It can be singular, untranslatable, and politically ontological or massively distributed, shared, and inescapable (De la Cadena and Blaser 2018).

Group→	1	2	3	4	5	6	7	8	9	10	11	12	13	14	15	16	17	18
Period 1	1 H																	2 He
2	3 Li	4 Be											5 B	6 C	7 N	8 O	9 F	10 Ne
3	11 Na	12 Mg											13 Al	14 Si	15 P	16 S	17 Cl	18 Ar
4	19 K	20 Ca	21 Sc	22 Ti	23 V	24 Cr	25 Mn	26 Fe	27 Co	28 Ni	29 Cu	30 Zn	31 Ga	32 Ge	33 As	34 Se	35 Br	36 Kr
5	37 Rb	38 Sr	39 Y	40 Zr	41 Nb	42 Mo	43 Tc	44 Ru	45 Rh	46 Pd	47 Ag	48 Cd	49 In	50 Sn	51 Sb	52 Te	53 I	54 Xe
6	55 Cs	56 Ba	57 La *	72 Hf	73 Ta	74 W	75 Re	76 Os	77 Ir	78 Pt	79 Au	80 Hg	81 Tl	82 Pb	83 Bi	84 Po	85 At	86 Rn
7	87 Fr	88 Ra	89 Ac *	104 Rf	105 Db	106 Sg	107 Bh	108 Hs	109 Mt	110 Ds	111 Rg	112 Cn	113 Nh	114 Fl	115 Mc	116 Lv	117 Ts	118 Og

*	58 Ce	59 Pr	60 Nd	61 Pm	62 Sm	63 Eu	64 Gd	65 Tb	66 Dy	67 Ho	68 Er	69 Tm	70 Yb	71 Lu
*	90 Th	91 Pa	92 U	93 Np	94 Pu	95 Am	96 Cm	97 Bk	98 Cf	99 Es	100 Fm	101 Md	102 No	103 Lr

FIGURE 6.1: The periodic table, as of 2020

Perhaps the most confident portrait of worldly elements today is offered by chemists, who, in the universal language of science, continue to chart the molecular building blocks of matter, offering on an ever-growing periodic table a technical description of all that is and that can be made (see figure 6.1). The configuration of the periodic table is open-ended, focusing on atomic structure and chemical properties, moving from those with fewest protons (hydrogen with one) to those with the most (currently oganesson with 118). The chart organizes the ninety-four known organic elements alongside the twenty-four elements that have been manufactured, for a current total count of 118 elements. Matter—both found and made—can be organized via these 118 categories, rendering an endless complexity of form out of a surprisingly limited number of basic elements. The periodic table is a remarkable portrait of both earthly matter and scientific achievement. In its open-ended form, the periodic table also promises that for every new element that is discovered or fabricated, a numbered box is always waiting, drawing anticipated novelty into an ever-expanding but perfectly ordered sequence of materiality. But who can live in such a perfectly ordered universe, acting *as if* the universe were so beautifully measured, calm, and color coded? What is the atomic weight of indifference or racism, pollution or madness?

In this chapter, I am not interested in all elemental matter but in a specific subcategory of artificial substance that has been stealthily remaking the earthly environment now for more than a half century. My focus is on those synthetic materials that, by virtue of industrial activity, mostly in the Global North, are now distributed literally at planetary scale. This variety of worldly essence is fundamentally a creation of a narrow group of people, generated by nuclear nationalism and petrochemical capitalism in a way that achieves planetary scale; not all have been equally involved in distributing these synthetic elements, even if everyone lives now in their violent wake (Sharpe 2016; Thomas 2019). The "elements" of concern in this chapter are therefore indexed to a specific mode of living consolidated in the mid-twentieth century that are foundational to a consumer economy and a nuclear superpowered state. These materials are found today literally across the entire earth system (in land, ice caps, oceans, air, and biosphere); thus, the materials I am concerned with constitute a new ontological condition on Earth, one achieved in less than five full generations of industrial life. These "elemental forms" have specific histories that are important to understand, connecting modes of production and consumption with new trajectories and temporalities of violence (see Murphy 2017; Shapiro 2015).

What I call ubiquitous elements are core achievements of nuclear nationalism and petrochemical capitalism, the perverse outcome of attempts to produce security, comfort, and convenience. Ubiquitous elements colonize a deep planetary future with the indelible marks of a specific socioeconomic order promoted enthusiastically by the United States after 1945. Thus, unlike many of the newest synthetic editions to the periodic chart—such as oganesson, which has existed for only a few microseconds as a laboratory experiment—these materials are, from a human point of view, permanent additions to the global environment. Ubiquity, here, is a register of a kind of totalizing territorial reach, while the elemental engages deep future time; together they reference invented molecular forms that, paradoxically, have both a short history (in existence for only decades) but that are now infrastructural to the composition of the earth. Therefore, ubiquitous elements raise profound philosophical questions about economy, about the modes of living that produce such complicated forms, and about the scope of human perception, deception, and self-awareness in a petrochemical-military industrial society (see Gray-Cosgrove, Liboiron and Lepawsky 2015; and Tsing et al. 2017).

A serious scientific investigation into ubiquitous elements has been undertaken by the Working Group on the Anthropocene (formally, the Subcommission on Quaternary Stratigraphy within the International

Union of Geological Sciences; see Zalasiewicz et al. 2019b). The Working Group was formed by geologists in response to overwhelming multidisciplinary evidence that industrial impacts on the environment have become so profound that they constitute a violent new aspect of the earth system, leading some to call for the designation of a new geological epoch—the "age of humans" or the Anthropocene (Crutzen and Stoermer 2000). This act of periodization would formally conclude the twelve thousand years of the Holocene (during which minor things such as language, agriculture, the internal combustion engine, smartphones, and nuclear weapons were invented); it would acknowledge the material force of industrial humanity as a geological agent going forward. The Anthropocene became a formal project of the Working Group in 2009, a serious scientific effort to assess geological conditions on Earth to see if a human-made signature exists on the right temporal and spatial scales to meet the strict professional standards of the International Commission on Stratigraphy. Staged cannily by members of the Working Group as a search for the "golden spike" of the Anthropocene (that is, the specific material formation and physical location that could anchor the new epoch), the inquiry provoked an unusually wide-ranging and heated conversation across the earth sciences, social sciences, and humanities about how to recognize human industrial impacts on the global environment (for a detailed account, see Bonneuil and Fressoz 2015).

As geologists scoured the earth searching for physical evidence of the Anthropocene, looking for a planetary-scale artificial material signal in the rock or ice that one could literally "hit with a hammer" (Zalasiewicz et al. 2019a, 3), scholars in the humanistic social sciences debated the implications of choosing one golden spike versus another for our understanding of contemporary conditions, with each option articulating a different diagnostic for when people began to materially and negatively change the natural order. The beginning of agriculture, the first steam engine, Columbus's trip to the New World, the plantation system, the invention of capitalism, and many more have received serious debate in a quite vibrant and often strange multidisciplinary conversation about violent origin points for the contemporary condition (see Moore 2016). This debate often merges the amplifying effects of global warming in the twenty-first century with the specific concerns of the Working Group about geology, creating a field of multiple competing anthropocenes, each offering a different diagnostic view on how the global environment became unhinged. This search for the golden spike has been highly productive, mobilizing social scientists and humanities scholars to rethink the temporal ordering of their disciplines, opening up deep time frames as relevant to their concerns while also engaging natural scientists in an unusual interdisciplinary conversation about historical, contemporary, and projected ecological conditions. But here, we should pause and ask: What are the elemental terms of a civilizational disorder that can generate planetary-scale environmental dangers and do so in multiple registers simultaneously without restraint? Where is such violence to be properly located on the periodic table? Via what specific molecular intensities has industrial modernity (with its profusion of technologies, petrochemical investments, and nuclear nationalisms) terraformed Earth, creating not only a new and volatile atmospheric chemistry but also cascading forms of ecological disruption (ocean rise, heat storms, droughts, and mass extinctions)?

Since the unit of the Anthropocene is the size of the planet itself, any viable candidate for the golden spike must operate at this ultimate scale and have a temporal form lasting on the order of at least tens of millennia. The official stratigraphic review requires the designation of a primary signature (a specific physical location on the planet for the placement of the golden spike), as well as a set of secondary signatures that are equally powerful in articulating a geological fact. Thus, any new geological epoch emerges out of a combined set of observations, each of which has to meet a specific technical judgment and be structured in relation to one another. The Working Group, while stirring up an unprecedented multidisciplinary debate, has endeavored to be apolitical in its judgment, focusing on the technical logics of the designation within geological science, on precisely what is in the strata of Earth. Nonetheless, the formal review of stratigraphic periodization in this case rides not simply on observation of the stratigraphic record but also on evaluating the vast range of planetary-scale industrial aftermaths created by people—namely, the unintended effects of an economy structured by petrochemical capitalism, nuclear nationalism, and globalized markets. Thus, the Anthropocene is unavoidably both political and diagnostic.

In their 2016 *Science* article, "The Anthropocene Is Functionally and Stratigraphically Distinct from the Holocene," members of the Working Group, led by Colin Waters, not only conclude that the mid-twentieth century is the moment when human activity achieves the necessary scale to demarcate a new geological epoch (thus ending the Holocene) but also offer a set of specific candidates for the golden spike. The Working Group attends here to the profound alternations in earthly conditions that occurred after World War II, when exponential shifts in travel, communication, and consumption index equally extreme changes in the global environment. This "great acceleration" (see McNeill and Engelke 2014) at mid-century includes the use of artificial compounds on a vast scale. Aluminum, concrete, plastics, and synthetic fibers become foundational materials for building cities, infrastructures, and consumer lifestyles after World War II, matched by rapid increases in greenhouse gases, biodiversity loss, ocean acidification, and global heating (see Steffen et al. 2015). The Working Group has investigated each of these artificial forms to see if any achieve the right scale and temporality for the golden spike (leaving unstated the resulting biological or ecological impacts which, by definition, are not metrics of geology), summarizing their conclusions on a chart of selected anthropogenic materials (see Waters et al. 2016, aad2622–3). As an index of modern industrial living, their chart (see figure 6.2) is remarkable as it assesses the proliferation of now omnipresent synthetic building materials, revealing the elemental forms of an increasingly urbanized global population that distributes toxins and greenhouse gases with each new building, road, or consumer item. Consider for a moment the built infrastructure of your current location: without concrete or aluminum or plastic, what would be left? Alternatively, consider the technofossils that are proliferating in your home space that will still be around in a distant future (see Mitman, Armiero, and Emmett 2018).

Of the universe of material possibilities, the Working Group has identified two anthropogenic signatures that stand out within the earth system: plutonium and plastic. I would like to consider each in turn and then contemplate the implications of these ubiquitous elements for worlding today. For what is at stake, in the judgment of the Working Group, is not only the technical progression of geology as a science—offering a more robust and careful measurement of stratigraphic history—but also the shifting status of people (really, capitalist consumers from the Global North) as terraformers. The driving force of radical environmental disruption is not all human beings but those embedded within a specific economic order organized by nuclear nationalism and petrochemical capitalism, involving those who not only inhabit Earth but increasingly intervene in its composition without much worry. As ubiquitous elements, plutonium and plastic are monumental, outrageous acts of

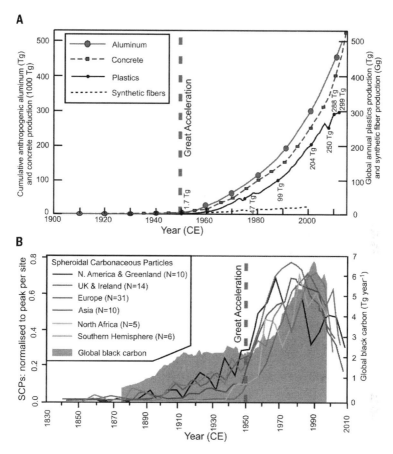

Fig. 2. The production of selected new anthropogenic materials. (A) Cumulative growth of manufactured aluminum in the surface environment [adapted from data in (*23*), assuming a recycling rate of 50%]; cumulative growth of production of concrete, assuming that most cement goes into concrete and that ~15% of average concrete mass is cement [from (*22*), derived from U.S. Geological Survey global cement production statistics]; annual growth of plastics production [from (*24*)]; and synthetic fibers production [from (*26*)]. **(B)** Global mid–20th century rise and late–20th century spike in SPCs, normalized to the peak value in each lake core [modified from (*32*)], and global black carbon from available annual fossil fuel consumption data for 1875–1999 CE (*30*). Numbers of lake cores for each region are indicated.

FIGURE 6.2: Chart of synthetic materials and the "great acceleration." From Waters et al. 2016.

coconstitution, offering lively and extremely perverse anthropogenic metrics of literal world making/world breaking on a planetary scale.

A PLUTONIUM PLANET

With the first atomic explosion in July 1945, followed quickly by the US strikes in Hiroshima and Nagasaki, many around the world felt a shift in the nature of reality, constituting a before-and-after moment for politics, ethics, and science with this hyperviolent start to a nuclear age. In Waters et al.'s (2016) study, the nuclear age figures loudly as an anthropogenic engine of planetary change, particularly in the form of atmospheric fallout from nuclear explosions. From 1945 to 2017, nuclear powers conducted 530 aboveground nuclear detonations (plus an additional 1,528 underground detonations), which distributed plutonium, cesium, and strontium globally (see Kimball 2020). The Working Group has examined ice-core samples and marine sediments around the world, identifying a clear plutonium spike that occurs in the period between 1952, with the first thermonuclear detonation, and 1963, with the signing of the Limited Test Ban Treaty (see figure 6.3). The treaty stopped nuclear testing in the atmosphere by the United States, the Soviet Union, and the United Kingdom; it was both the start of nuclear arms control and the first international environmental treaty. This international effort to reduce radioactive fallout, however, allowed nuclear testing to continue underground into the 1990s, ultimately revealing a choice among nuclear powers to support a hyper-dangerous industry over and against the earthly environment, which became permanently marked not only with radioactive fallout but also with underground test craters, nuclear waste, and zones of massive contamination (see Masco 2020).

In considering the global distribution of different long-lived fallout radionuclides, the Working Group notes that cesium-14 has some natural sources, making it less viable for the Anthropocene designation than plutonium, which stands out today as the clearest purely artificial signal in the stratigraphy of the planet:

> Pu-239, with its long half-life (24,110 years), low solubility, and high particle reactivity, particularly in marine sediments may be the most suitable radioisotope for marking the start of the Anthropocene. The appearance of a Pu-239 fallout signature in 1951 CE,

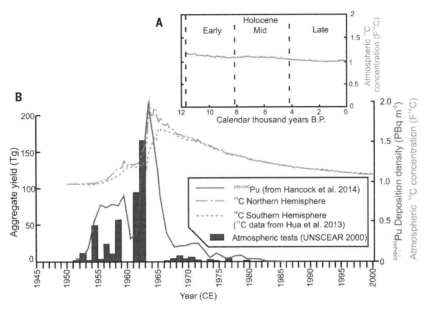

Fig. 4. Radiogenic fallout signals as a marker of the Anthropocene. (A) Age-corrected atmospheric ^{14}C concentration (F[14]C) based on the IntCal13 curve, before nuclear testing (62). (B) The atmospheric concentration of [14]C (F[14]C) (63) and [239+240]Pu (64) radiogenic fallout from nuclear weapons testing (PBq, petabecquerel), plotted against annual aggregate atmospheric weapons test yields (60).

FIGURE 6.3: Spike in plutonium and cesium from nuclear detonations. From Waters et al. 2016.

peaking in 1963–64 CE will be identifiable in sediments and ice for the next 100,000 years. (Waters et al. 2016: aad2622–5)

Identifiable for the next 100,000 years. For the Working Group, plutonium fallout from 1952 to 1964 offers an artificial signature at the right scale and planetary distribution for the Anthropocene designation. This calibration offers a shocking specificity to the Anthropocene—for example, in 1952 there were only two nuclear powers, the United States and the Soviet Union. And this periodization elevates not the atomic age (which would start on July 16, 1945, with the first atomic detonation in New Mexico) but the thermonuclear age to a formal geological periodization (recorded to the precise second when the first thermonuclear detonation occurred on November 1, 1952, at 7:15 a.m. in the Marshall Islands). For the Working Group, the bomb is important not as a new means of war, nor as a new mode of living (see Masco 2020). The materiality of nuclear nationalism matters to geology today as an industrial aftermath, constituting a plutonium distribution at planetary scale that is so highly visible in the stratigraphic record that it could be found by a visitor from another planet. For this reason, the Working Group has formally proposed that plutonium from early Cold War nuclear explosions, the most explicit and distinct artificial signature in the stratigraphic record, be the lead anchor for the Anthropocene designation, transforming the 24,110-year half-life of the element into the primary argument for a new conceptualization of earthly history (for the latest updates on the formal review process, see Anthropocene Working Group n.d.).

Plutonium occurs only in trace amounts via natural process but is now ubiquitous as an artificial substance on Earth—a key achievement of twentieth-century nuclear nationalism. Formally, plutonium is element 93 on the periodic table. It was discovered in 1940 by Glenn Seaborg, Edwin McMillan, J. W. Kennedy, and A. C. Wahl at the University of California, Berkeley. Seaborg wrote the first paper on the element, which was officially censored as he took up plutonium production at the University of Chicago for the Manhattan Project. The Hanford Site in Washington State was soon established to start mass production of plutonium for the war effort, which it did through the end of the Cold War, eventually producing 111 tons of the stuff for the United States. About ten pounds of plutonium is needed to produce a bomb, and it takes a substantial industrial infrastructure to convert uranium to plutonium. There are now about five hundred tons of plutonium stored on earth, located across production sites, reactors, and weapons programs. As a dangerous material subject to industrial accidents and explosions and as the key to building the ultimate weapon of mass destruction, plutonium is highly fetishized, tracked, and worried about. There is a global program to monitor plutonium and multiple international efforts to keep track of each molecule of bomb-grade fissile material under rubrics of collective defense (see Feiveson et al. 2014). In fact, one can think about the plutonium economy as linking people, ecologies, cultures, and economies in highly novel, and still emerging, configurations (see Masco 2006; Brown 2013).

Plutonium continues to produce elemental emotions today—from fear and anger to awe and, for a few, something like love. In its glowing form, warm to the touch, it is an emblematic example of the coconstitution of society and nature—of the ways that technoscientific revolution is world making and world breaking all at once. Worldly elements can be infused with affect, desire, misrecognition, and psychosis. In the post–Cold War 1990s, I collected stories from weapons scientists about plutonium. It was often referred to as bizarre, fascinating, perplexing, counterintuitive in terms of how it interacted with the world, hot to the touch, unpredictable over time, and scary to work with as both an unstable element and a poison. Two top plutonium scientists at Los Alamos wrote about the element, and its post–Cold War condition, this way:

> Like other reactive materials, plutonium ages with time. In moist air, it "rusts" much more profusely than iron, and when exposed to other atmospheric environments, it will react to form several surface-corrosion products. In other words, plutonium ages from the outside in. What makes plutonium really special, however, is that is also ages from the inside out. As a result of its radioactive nature, it relentlessly undergoes self-irradiation damage through its volume. Consequently, nature's most unusual element becomes even more complex as it ages. In the past, we were resigned to keeping plutonium from self-destructing—at least for two or three decades. Today, we are intensely interested in extending its storage life for many more decades, preferably as much as a century (Hecker and Martz 2000, 238).

Nature's most unusual element. Here, we might pause to consider what kind of nature is being evoked here, as the only naturally occurring plutonium was produced in what is now Gabon over two billion years ago, when a natural reactor was formed underground converting some uranium deposits to plutonium. This form of plutonium has long since dissipated, leaving only a chemical trace of an extraordinary, and extraordinarily ancient, geological event. The earthly presence of plutonium-239, however, has been produced with great effort by a few nation-states since 1945 and is a singular index of nuclear militarism, used primarily in making weapons and occasionally as a power source for satellites and spacecraft.

The ongoing interactions between plutonium and the environment are hugely complex and, because of the material's strange phase states and aging properties, are as unpredictable as the chemical composition of the element itself. To be precise: the value and danger of plutonium are precisely its instability. Much of the US supply of plutonium was made during the mid-twentieth century at the Hanford Site (although there is currently an effort to establish a new plutonium production capacity in the United States as part of an effort to rebuild the nuclear arsenal in the twenty-first century). Plutonium leaking from gigantic vats of liquid waste at Hanford is predicted to reach the Columbia River at some point over the next one thousand years—a slow-moving but inevitable legacy of the Cold War (see Cram 2015). Similar environmental problems exist in the plutonium complex of the former Soviet Union, representing a permanent legacy of that sixty-nine-year nuclear national experiment (see Brown 2013).

Over 100,000 Cold War–era employees of the US nuclear complex have now applied for compensation for on-the-job medical injury under the highly limited terms allowed by the US Congress. There are many other populations around the world affected by fallout (from military strikes, nuclear test programs, nuclear accidents, and nuclear waste storage) that have no comparable compensation program. The material force of plutonium since 1945 is one that crosses and differentially links ecologies, workers, victims, geopolitical agendas, enemy configurations, and technoscientific revolutions. Plutonium has been central to world making in terms of geopolitics, international institutions, and configurations of national power but also world breaking in terms of the contamination of specific bodies, populations, and ecosystems, and, as the Anthropocene debate reiterates, it continues to generate new planetary-scale dynamics.

Plutonium-239, which is the key form for military industrial applications, has a half-life of just over twenty-four thousand years or twice the length of the Holocene, which is why it is of principal concern not only for nuclear states seeking to extend their commitment to the bomb in the twenty-first century but also as a geological marker for the earth sciences. The centrality of plutonium to the Anthropocene debates should not be surprising, as many aspects of earth system science were first enabled by following the distribution of radionuclides from nuclear detonations as they moved across air, ocean, land, and ice in the mid-twentieth century (Masco 2010). This early moment in the nuclear age links military science to environmental sciences in a foundational way, one that can also be located in the career trajectories of many of the central figures in the Anthropocene review.

For members of the Working Group, plutonium works as the primary anthropogenic signal because it is so highly visible in the stratigraphic record—artificial, widely distributed, and long-lived. Plutonium also makes sense to many because it is the elemental form of a new species of mass violence—nuclear war—and because global nuclear arsenals maintain the capacity to eliminate life on Earth. The plutonium fallout from the mid-twentieth century will likely outlive any current understanding of the Cold War, or of a nuclear age, and quite possibly ideas about the Anthropocene as well. But what these material observations miss is the continuing power of plutonium as a psychosocial force. After 1945, plutonium is at the center of a world system, the central material in an arms race, a motivating rationale in the building out of international law and nonproliferation regimes, a source of terror for officials and citizens alike, and a molecular force that can be tracked across technoscientific regimes, industrial economies, national security states, and the nightmares that organize specific social orders and individual psyches. The worlding work of plutonium, in other words, matters well beyond questions of periodization. As an extremely rare material, made at great expense, which is highly protected, surveilled, and sought after, plutonium continues to produce intense emotions; it is also the definitional opposite of the Working Group's second primary candidate for the golden spike: plastic.

LIVING IN A PLASTISPHERE

Roland Barthes ([1957] 1995) wrote a short essay on plastic more than six decades ago that is eerie from the perspective of the current Anthropocene debates. Identifying a material that is "ubiquity made visible," a substance of "infinite transformation," Barthes sees plastic as the first form in human history that strives to be mundane rather than rare, everyday rather than special, useful rather than filled with aura. He argues that plastic:

> is the first magical substance which consents to be prosaic. But it is precisely because this prosaic character is a triumphant reason for its existence: for the first time, artifice aims at something common, not rare. . . . A luxurious object is still of this earth, it still recalls, albeit in a precious mode, its mineral or animal origin, the natural theme of which it is but one actualization. Plastic is wholly swallowed up in the fact of being used: ultimately, objects will be invented for the sole pleasure of using them. The hierarchy of substances is abolished: a single one replaces them all; the whole world can be plasticized, and even life itself since, we are told they are beginning to make plastic aortas. (63)

The whole world can be plasticized. In light of the Working Group's 2016 assessment, Barthes's essay is nothing short of prophetic. Barthes sees in the mid-1950s an emerging world order devoted to convenience and consumerism, promising a steady colonization of everyday life by plastic. The global success of plastic is beyond all measure and has become an everyday nightmare: researchers have found plastic in the deepest part of the Pacific Ocean and at the top of mountains; it is not only in landfills but can be found floating in the air as microplastic. By the twenty-first century, one could locate plastic inside nearly every organism and body of water on the planet (Liboiron 2015). So how on earth did such a mundane material, designed and deployed for convenience, become a planetary force?

Plutonium may stand out to geologists in the stratigraphic record because it is a singular signal of pure artificiality, a human-generated presence that can be found widely but also dated precisely. Plastic, on the other hand, consists of a variety of molecules, synthesized mostly out of petrochemical processes, that can be formed into any shape and are generated by many different corporations around the world. Plastics are polymers, constituting a range of chemical arrangements, including polycarbonates, polyesters, polyethylenes, polypropylenes, polystyrenes, and polyurethanes—used respectively in eyeglasses, clothing, water bottles, straws, compact discs, and car parts, to name but a few. Many plastics are designed as throwaways, as items that are utilitarian and convenient but are not imagined as long-term forms (Gabrys, Hawkins, and Michael 2013). The synthetic materials that make up plastics, however, are extremely long-lived, all but eternal. Geologists are now finding—and predicting the discovery of vastly more—"technofossils" around the world. These plasticized objects will be found in the geology of the earth for millennia to come (Zalasiewicz et al. 2016). Compact discs, phones, water bottles: the list is endless and is a mirror to the forms of everyday life in a petrochemically based consumer society.

Think of your cell phone or the plastic fork from lunch or your toothbrush—not as it is today but as it will be one hundred or a thousand years from now, when you are long gone. Plastic inhabits a diabolical contradiction, in that its value is as a material used in packaging and for mass-produced items primarily for convenience, but in each and every instance plastic is also a potentially permanent contribution to the future geological makeup of Earth. The chemical properties that make plastic cheap and fun to work with also make it eternal, so much so that many current assessments of the amount of plastic on Earth envision its future planetary presence as being on the scale of the other spheres—biosphere, atmosphere, hydrosphere, lithosphere: in other words, a "plastisphere" (see Liboiron 2014). Indeed, the World Economic Forum (2016, 7) predicts an ever-accelerating reliance on plastic in the twenty-first century and huge environmental effects, including a scenario in which the plastic in the oceans soon outweighs the fish. Other studies predict an accelerating use of plastic in the twenty-first century, more than doubling the volume of plastic waste in the environment by 2050 to over twelve billion tons and raising a series of questions about how the biosphere—literally all organisms—will be affected by plastic saturation (see Cox et al. 2019).

Here is how the Working Group assesses plastics as part of the Anthropocene review:

> The manufacture of new organic polymers (plastics), which were initially developed in the early 1900s, rapidly grew from the 1950s to an annual production of about ~300 Tg in 2013, comparable to the present human biomass. Plastics spread rapidly via rivers into lakes, and they are now also widespread in both shallow- and deep-water marine sediments as macroscopic fragments and as virtually ubiquitous microplastic particles (microbeads, 'nurdles,' and fibers), which are dispersed by both physical and biological processes. The decay resistance and chemistry of most plastics suggest that they will leave identifiable fossil and geochemical records. (Waters et al. 2016, aad2622-3)

Indelible records. Plastics do not, for the most part, decompose. Indeed, projections about the end of plastic currently rely on a perverse assumption—namely, that given the amount of plastic in the environment, a microorganism will inevitably evolve that can feed on plastic and begin to recycle it (Davis 2015, 233).

As Max Liboiron (2015, 3) puts it, "Plastic pollution currently exceeds the ability of traditional scientific methods to explain its fate and transport, as well as its persistence and effects. For example, every human and animal body tested in the past decade contains chemicals that leach from plastics, meaning that it is no longer possible to establish uncontaminated control groups for experimental designs in scientific research on the effects of plastics." In other words, we are all now part of the plastisphere, and plastic has become a new milieu for life on Earth. Some plasticizers and additives used in making plastic have been identified as endocrine disrupters, which means that the plastisphere is not just a geological problem or ecological surround; it challenges reproduction for humans and other species (Davis 2015). Plastic is thus a creation of people that also intrudes upon the biological makeup and futurity of all living beings. A plastic world is one that is designed for human convenience, converting petrochemical reserves into an infinite variety of forms and uses. It also, as Barthes predicted, colonizes land, water, and bodies, becoming an anthropogenic force par excellence. If the world is now wrapped in plastic, making the polymer elemental to earthly conditions in the twenty-first century, how does one account for the passions, blindness, disregard, and modes of self-absorption that enabled it to be so? How should one think about the transformation of everyday activities—eating, drinking, communicating, traveling—into a cumulative collective environmental assault, one that has in less than eight decades literally infiltrated earthly conditions?

THE FETISH AND THE DISCARDED

The energetic global search for evidence in support of the Anthropocene has produced an astonishing insight into industrial life in the Global North since 1945: both the most fetishized of materials (plutonium) and the most mundane (plastic) now operate on a planetary scale, each producing a set of ecological effects that matter and continue to unfold but that were neither intended nor desired. What does it mean that scarcity (plutonium) and superabundance (plastic) matter equally to the Anthropocene? How is it that a plutonium planet and a plastisphere reinforce each other as material indexes of a specific mode of living, as nuclear nationalism and petrochemical capitalism have, in the day-to-day practices across only a few generations, turned the biosphere itself into an externality for industrialism, a living space sacrificed repeatedly for momentary security and convenience?

The turn to elemental matter is a vital project today precisely because it can reveal contradictions which are so embedded in existing institutions as to be rendered natural, ideologically positioned as unavoidable, or essential to assumptions about technoscientific or economic "progress." The Working Group has revealed that regardless of whether a material is mundane or rare, dangerous in small amounts or only at massive scale, heavily policed and the central concern of a security-state apparatus or multiply sourced and essential to global commerce, too hot to

Eonothem / Eon	Erathem / Era	System / Period	Series / Epoch	Stage / Age	GSSP	numerical age (Ma)
Phanerozoic	Cenozoic	Quaternary	Holocene	U/L Meghalayan		present 0.0042
				M Northgrippian		0.0082
				L/E Greenlandian		0.0117
			Pleistocene	U/L Upper		0.129
				M Chibanian		0.774
				L/E Calabrian		1.80
				Gelasian		2.58
		Neogene	Pliocene	Piacenzian		3.600
				Zanclean		5.333
			Miocene	Messinian		7.246
				Tortonian		11.63
				Serravallian		13.82
				Langhian		15.97
				Burdigalian		20.44
				Aquitanian		23.03
		Paleogene	Oligocene	Chattian		27.82
				Rupelian		33.9
			Eocene	Priabonian		37.71
				Bartonian		41.2
				Lutetian		47.8
				Ypresian		56.0
			Paleocene	Thanetian		59.2
				Selandian		61.6
				Danian		66.0
	Mesozoic	Cretaceous	Upper	Maastrichtian		72.1 ±0.2
				Campanian		83.6 ±0.2
				Santonian		86.3 ±0.5
				Coniacian		89.8 ±0.3
				Turonian		93.9
				Cenomanian		100.5
			Lower	Albian		~ 113.0
				Aptian		~ 125.0
				Barremian		~ 129.4
				Hauterivian		~ 132.6
				Valanginian		~ 139.8
				Berriasian		~ 145.0

FIGURE 6.4: The future location of the Anthropocene on top of the Holocene. Courtesy of International Commission on Stratigraphy, www.stratigraphy.org.

touch or a throwaway object, a planetary ecological effect of multimillennial duration can be achieved. An assessment of elemental matter today, therefore, should not stop with a forensic account of contemporary ecological conditions; it should also consider the lifeworlds that produce such profound ecological effects, whether in the name of national security or for the sake of consumer convenience. What can one now say about the social orders that can produce violence on so vast a scale, creating planetary-scale problems without intending to do so or not stopping once that violence is understood to be occurring? And specifically, what should be the second sentence after geologists formally enter plutonium and plastic into the geological deep time of the planet as the Anthropocene?

The International Chronostratigraphic Chart (ICC) (see figure 6.4) will very likely soon receive a new top layer with the Anthropocene designation, offering a revised depiction of earthly time, extending its perfected color-coded system of eons, eras, periods, epochs, and ages. Like the periodic table, the ICC depicts a rational world of ordered relations and anticipates future additions, leaving open the question of what might come after the Anthropocene. But plutonium and plastic, as we have seen, are deeply irrational forms as well as technoscientific achievements, consolidating fears and desires, revealing choices and priorities that have consistently rendered a livable earth an afterthought. As now-ubiquitous elements of the earth system, as both manufactured and eternal, plutonium and plastic provoke questions not only about the scope and coconstitution of earthly conditions. These materials will continue to transform the collective milieu in perpetuity, confirming and reconfirming in unforeseeable ways the relentless anthropogenic force of twentieth-century nuclear nationalism and petrochemical capitalism. As elemental forms, plutonium and plastic will reveal and trouble, as well as shape and derange, the potentials, qualities, and ambitions for what constitutes life in an increasingly artificial world.

REFERENCES

Anthropocene Working Group. n.d. Subcommission on Quarternary Stratigraphy. Accessed March 15, 2021. http://quaternary.stratigraphy.org/working -groups anthropocene/.

Barthes, Roland. (1957) 1995. "Plastics." *Daidalos* 56:61–63.

Bonneuil, Christophe, and Jean-Baptiste Fressoz. 2015. *The Shock of the Anthropocene*. New York: Verso.

Brown, Kate. 2013. *Plutopia: Nuclear Families, Atomic Cities, and the Great Soviet and American Plutonium Disasters*. Oxford: Oxford University Press.

Cox, Kieran D., Garth A. Covernton, Hailey L. Davies, John F. Dower, Francis Juanes, and Sara E. Dudas. 2019. "Human Consumption of Microplastics." *Environmental Science and Technology* 53 (12): 7068–74.

Cram, Shannon. 2015. "Becoming Jane: The Making and Unmaking of Hanford's Nuclear Body." *Environment and Planning D: Society and Space* 33 (5): 796–812.

Crutzen, Paul J., and Eugene F. Stoermer. 2000. "The 'Anthropocene.'" *Global Change Newsletter* 41:17–18.

Davis, Heather. 2015. "Toxic Progeny: The Plastisphere and Other Queer Futures." *philoSOPHIA* 5 (2): 231–50.

De la Cadena, Marisol, and Mario Blaser, eds. 2018. *A World of Many Worlds*. Durham, NC: Duke University Press.

Feiveson, Harold, Alexander Glaser, Zia Mian, and Frank N. von Hippel. 2014. *Unmaking the Bomb*. Cambridge, MA: MIT Press.

Gabrys, Jennifer, Gay Hawkins, and Mike Michael, eds. 2013. *Accumulation: The Material Politics of Plastic*. New York: Routledge.

Gray-Cosgrove, Carmella, Max Liboiron, and Josh Lepawsky. 2015. "The Challenges of Temporality to Depollution and Remediation." *SAPIENS* 8 (1): 1–10.

Hecker, Siegfried, and Joseph C. Martz. 2000. "Aging of Plutonium and Its Alloys." *Los Alamos Science* 26:238–43.

Kimball, Daryl. 2020. "The Nuclear Testing Tally." Website of the Arms Control Association. Last reviewed July 2020. https://www.armscontrol.org /factsheets/nucleartesttally.

Leslie, Esther. 2005. *Synthetic Worlds: Nature, Art and the Chemical Industry*. London: Reaktion.

Liboiron, Max. 2014. "The Plastisphere." In *The Petroleum Manga: A Project by Marina Zurkow*, edited by Valerie Vogrin and Marina Zurkow, 24. Brooklyn: Peanut Books.

Liboiron, Max. 2015. "Redefining Pollution and Action: The Matter of Plastics." *Journal of Material Culture* 21 (1): 1–24.

Liboiron, Max. 2021. *Pollution Is Colonialism*. Durham, NC: Duke University Press.

Masco, Joseph. 2006. *The Nuclear Borderlands: The Manhattan Project in Post–Cold War New Mexico*. Princeton, NJ: Princeton University Press.

Masco, Joseph. 2010. "Bad Weather: On Planetary Crisis." *Social Studies of Science* 40 (1): 7–40.

Masco, Joseph. 2020. *The Age of Fallout and Other Episodes in Radioactive World-Making*. Durham, NC: Duke University Press.

McNeill, J. R., and Peter Engelke. 2014. *The Great Acceleration: An Environmental History of the Anthropocene since 1945*. Cambridge, MA: Belknap.

Mitman, Gregg, Marco Armiero, and Robert S. Emmett, eds. 2018. *Future

Remains: A Cabinet of Curiosities for the Anthropocene. Chicago: University of Chicago Press.

Moore, Jason W., ed. 2016. *Anthropocene or Capitalocene? Nature, History and the Crisis of Capitalism*. Oakland, CA: PM Press.

Murphy, Michelle. 2017. "Alterlife and Decolonial Chemical Relations." *Cultural Anthropology* 32 (4): 494–503.

Shapiro, Nicholas. 2015. "Attuning to the Chemosphere: Domestic Formaldehyde, Bodily Reasoning, and the Chemical Sublime." *Culture Anthropology* 30 (3): 368–93.

Sharpe, Christina. 2016. *In the Wake: On Blackness and Being*. Durham, NC: Duke University Press.

Steffen, Will, Wendy Broadgate, Lisa Deutsch, Owen Gaffney, and Cornelia Ludwig. 2015. "The Trajectory of the Anthropocene: The Great Acceleration." *Anthropocene Review* 2 (1): 81–98.

Stengers, Isabelle. 2015. *In Catastrophic Times: Resisting the Coming Barbarism*. London: Open Humanities.

Thomas, Deborah. 2019. *Political Life in the Wake of the Plantation: Sovereignty, Witnessing, Repair*. Durham, NC: Duke University Press.

Tsing, Anna, Heather Swanson, Elaine Gan, and Nils Bubandt, eds. 2017. *Arts of Living on a Damaged Planet*. Minneapolis: University of Minnesota Press.

Waters, Colin N., Jan Zalasiewicz, Colin Summerhayes, Anthony D. Barnosky, Clement Poirier, Agnieszka Galuszka, Alejandro Cearreta, et al. 2016. "The Anthropocene Is Functionally and Stratigraphically Distinct from the Holocene." *Science* 351 (6269): aad2622-1–10.

World Economic Forum. 2016. *The New Plastics Economy: Rethinking the Future of Plastics*. Geneva: World Economic Forum.

Zalasiewicz, Jan, Colin N. Waters, Juliana A. Ivar do Sul, Patricia L. Corcoran, Anthony D. Barnosky, Alejandro Cearreta, Matt Edgeworth, et al. 2016. "The Geological Cycle of Plastics and Their Use as a Stratigraphic Indicator of the Anthropocene." *Anthropocene* 13:4–17.

Zalasiewicz, Jan, Colin N. Waters, Mark Williams, Colin P. Summerhayes, Martin J. Head, and Reinhold Leinfelder. 2019a. "A General Introduction to the Anthropocene." In *The Anthropocene as a Geological Time Unit: A Guide to the Scientific Evidence and Current Debate*, ed. Jan Zalasiewicz et al.

Zalasiewicz, Jan, Colin N. Waters, Mark Williams, and Colin P. Summerhayes, eds. 2019b. *The Anthropocene as a Geological Time Unit: A Guide to the Scientific Evidence and Current Debate*. Cambridge: Cambridge University Press.

7

TILTING AT WINDMILLS

Patrick Bresnihan

> Before 19th-century thermodynamics—which came out
> of the development of fossil-fuelled steam engines and
> electric motors and batteries—nobody talked about en-
> ergy at all in its current sense. People talked about horses,
> fire, trade winds, lightning, the ripening of wheat and so
> forth. Each was connected less with a common "energy"
> than with ploughing, cooking, sailing, eating and so on,
> with all the limits to interchangeability and accumula-
> tion that the existence of such disparate meshes of rela-
> tionships imply. In some contexts, the actions and beings
> involved were even seen as having individual personali-
> ties, dignities or rights.
> —Larry Lohmann and Nicholas Hildyard,
>   "Energy, Work, and Finance"

In the west of Ireland, the wind blows most of the time. What hits the
land has already traveled across the Atlantic, warming up and collect-
ing moisture along the way. Besides bringing rain and unsettled weather
patterns, the wind leaves lasting impressions on the trees, dunes, buildings, and bodies of the people who live there. I have a clear image of three
whitethorn trees growing in the field in front of my childhood house.

They are bent to a forty-five-degree angle, their branches on one side resting on the ground. They sweep from left to right as you look at them out of the kitchen window, a clean line that draws your eye in a neat movement. There is a German word for these trees, *krummholz*—literally "crooked wood," meaning stunted and wind-blasted trees. Whitethorn is one of the few trees that can withstand the blast of the wind; most trees do not grow above the protective heights of walls or buildings due to "wind burn," the lash of salt carried by the Atlantic-fashioned wind, which also burns the faces of those exposed to it. The muffled sound of the wind outside, the whistle through a chimney, or the ticking of rope on boat masts becomes so familiar that when a calm day does come, the quiet can be deafening, leaving people a little disoriented. There is even a saying in the west of Ireland that when the wind stops blowing, everyone falls down.

Wind is not air but the movement of air. Wind is caused as air moves from areas of high pressure to areas of low pressure and is thus a manifestation of complex atmospheric dynamics that connect the sun's radiation, the shape of the land or sea, and the rotation of the earth. Wind is observable in ways air is not. It takes form through its action on things—the crooked trees, the ticking rope. In 1805 Francis Beaufort designed the first wind scale based on his observations of the wind acting on the sea: at 0, "the sea is like a mirror"; at 12, "the air is filled with foam and spray." In his book *Where the Wild Winds Are*, the nature writer Nick Hunt (2017) crosses Europe by foot to follow four of Europe's winds: the Helm, the Bora, the Foehn, and the Mistral. What he finds are not just winds acting on his body but entire landscapes, ecologies, and cultures that have been shaped by these unseen forces. Wind as element is not discrete; it is ecological, always composed of and with others beyond the aims and control of humans.

In Greek mythology, the gods of the wind were the Anemoi, from *anima*, "breath" or "soul." Nick Hunt describes his journey after the winds as a journey into animism: an understanding of the world as a living, breathing body. The winds he finds bring seasonal change as well as madness and despair. Winds have always been both a curse and a blessing bestowed by the gods—in myths and legends, the winds frequently cause trouble or create chance encounters by sending the protagonists off course (Peters 2012). In Christian and pagan imaginaries, wind is also associated with life and death, a force outside human control and thus unpredictable. The romantic poetics of wind borrows from these traditions, evoking the uncanny powers of nature, qualities that unsettled confidence in Enlightenment reason (Abrams 1957): it was the wind that stirred the soul, that conjured an "unseen presence." Drawing on classical mythology, Coleridge's "Eolian Harp" speculated that all animated nature was nothing but organic wind harps, diversely framed, through which sweeps "one intellectual breeze, / At once the Soul of each, and God of all" (Abrams 1957, 123).

Today, wind figures again as life taker and life giver. With climate change, wind unites more intensely and frequently with the elements of water and fire to generate superstorms and uncontrollable wildfires. As the power of these irregular winds becomes more pronounced, wind is simultaneously championed as the energetic force that can sustain life as we know it after the end of fossil fuels. In contrast to the chaotic winds of climate change, wind as energy source evokes imaginaries of technological mastery. In a remarkable short film produced in 2016 by the Irish Wind Energy Association, a lobby group for private wind energy developers, a recording of John F. Kennedy's speech to the Irish parliament in 1963 is cut with emotive music and images of contemporary Irish society. As Kennedy describes Ireland's successful journey toward becoming a modern, developed country since independence in 1922, an isolated, rural house is suddenly lit up, followed by Dublin's docklands (also a financial services center). In typical Kennedy style, the speech goes on to call for more to be done, for bigger dreams and bolder futures. Reference to global responsibilities coincides with an image of power plants pumping out CO_2 emissions. The eyes of Irish children reflect the crashing of waves and a landscape of electricity pylons. At last we see a horizon of wind turbines as a young boy plays in the foreground. The film ends by asking why Ireland imports 85 percent of its energy when it is surrounded by an energy resource that could grant energy independence. Kennedy's prophetic speech in 1963, it turns out, was about Ireland's new struggle for (energy) independence as part of the bigger, global mission to shift toward clean, green, renewable energy. Combining narratives of progress (control of nature and the technological sublime) with the environmental agenda, the result is a compelling "salvation" story that can be hard to challenge (Weston 2012).

Today's energy futures are being shaped by discourses of climate change and the cumulative, previously uncounted, limits of fossil fuel energy systems. The ambient, airy quality of wind is easy to contrast with the dirty, material qualities of coal and oil. Wind energy fits neatly within a more general shift in "green" economic thought toward the idea of nature as service provider, a flow of ecosystem or atmospheric services that promises abundance without the negative "externalities" of fossil fuels. The common good remains wedded to economic growth and progress but bathed in a green hue that extends to a global scale. Just as coal power was, to begin with, a speculative endeavor that relied on new fields of expertise, from geology to engineering, to justify the replacement of cheaper, cleaner forms of energy (water, wind, animal or human labor), so too have the past thirty years witnessed the rendering of wind as a "green" economic/energy reality through the development of new technoscientific discourses and infrastructures.

Beneath the bold rhetoric of "win-win" solutions for people and environment are familiar tensions and disputes over who stands to benefit (and who to lose out) from the clean energy transition. In Ireland, the expansion of wind energy capacity has been met with significant local opposition to the siting and development of not only wind farms but the electricity pylons and battery farms that are key elements of Ireland's emerging new energy infrastructure. Disputes over wind energy infrastructures in Ireland (and elsewhere) are often represented in the national (urban-based) media as "NIMBY" (not in my backyard) politics—local interests blocking projects for the greater good. A more critical position, adopted by a spectrum of environmental NGOs, political parties, campaign groups, academics, and activists, argues that local communities would get behind wind energy projects if they could benefit from them and have a greater role in their planning. Here, criticism is aimed at *how* wind energy is being developed—namely, through reliance on large, corporate wind developers and investors. Contradictions arise as the needs (profitable returns) of global energy companies, shareholders, and investors are prioritized over the needs of local communities/landscapes and the energy-consuming public more generally (Klein 2015; Schwartzman 2015). The extension of familiar logics of capitalist accumulation has led to these industrial wind energy developments being described as "aeolian extractivism" (Boyer 2011).

What does it mean to put the word *energy* after wind? How does it transform this element in motion, not just symbolically but through the complex elaboration of material practices and infrastructures designed to put "it" to work?[1] This chapter takes the invitation to think elementally with wind to interrogate and unsettle the extractive logics that seek to render wind as universal work/energy. Extractive logics, to paraphrase Kathryn Yusoff, involve both the making passive of certain bodies and territories (awaiting extraction and possessing of properties) and their activation through the mastery of white men (Yusoff 2018). In her powerful text *A Billion Black Anthropocenes or None*, Yusoff traces the entwined histories of geology, slavery, and colonialism. She describes the indifferent logics of extraction as being motivated by a desire for inhuman properties—in black and brown bodies as well as in the earth—and inseparable from the history of settler-colonialism in the New World.[2] The "problem" of geology and the "problem" of slavery were inextricably linked to extraction, she argues, setting in motion a history of accumulation and violence that is more often narrated as though it began in the European nineteenth century with industrialization and the mining of coal (see Malm 2016).

Identified as *terra nullius*, the colonial frontier is first voided of any preexisting attachments and epistemic practices; frontiers are always imagined (and constructed) as sites of "bountiful emptiness," "empty but full" (Bridge 2001). Subsequently, as the frontier is mapped, categorized, and ordered, it becomes productive, harnessed to the interests of the colonizing power. It is harder to imagine these extractive, frontier logics relating to wind because wind is not body or territory. Yet that does not mean it cannot be subjected to logics of extraction. Wind is being reduced to certain inhuman properties—namely, potential energy—that can then be activated through the technical mastery of smart energy grids. This process is not sudden or spectacular but involves new fields of technoscientific expertise, the expansion of physical infrastructures, the geological extraction of minerals, and the repurposing of entire landscapes. Wind may not have body, but its capture, circulation, and use within energy-intensive capitalism does.

The motivation for this chapter follows others in this volume both in calling to "dismantle" colonial epistemic practices and relations to matter and in searching for alternative forms of experimental practice. This is necessary when the catastrophic history of the past five hundred years continues to be cast as a long series of unintended side effects of Western progress. This "cosy humanism" (Yusoff 2018) presents environmental problems as a series of "externalities" to be accounted for and managed. Integral to this positive green agenda is, as Papadopoulos identifies in this volume, a belief in "like for like" substitution. Materials, energy sources, labor, technologies, and even behaviors can be swapped in and out as though globally composed, place-specific ecologies were a single, vast accounting ledger. Ongoing efforts to harness wind as renewable energy is one example of this. Rather than an innocent move toward a cleaner, more sustainable future, however, this process can be understood as a continuation of the extractive logics of the past that have not served the majority well.

THE FIRST COMMERCIAL WIND FARM in Ireland began generating electricity in 1992, the year the Rio de Janeiro Earth Summit launched "sustainable development." It is fitting, then, that the first turbines were erected on a site previously designated for fossil fuel extraction.[3] It was not until the turn of the new century, however, that these early entrepreneurial ventures were given new impetus from the state. At a European level, the adoption of the Treaty of Lisbon in 2000 set out an ambitious ten-year strategy based on technological innovation and environmental renewal, including the shift away from fossil fuels. In 2002, the European Union (EU) and all its member states ratified the Kyoto Protocol, pledging the EU to a reduction in greenhouse gas emissions. The same year, the Irish government established the Sustainable Energy Authority of Ireland (SEAI), designed to transform Ireland into a society based on sustainable energy systems. A key part of its mission was to ramp up renewable energy generation with the aim of decarbonizing electricity, driving in turn the electrification of transport and heating.

One of the first acts of the new sustainable energy authority was to produce the first Irish wind atlas. The wind atlas displays wind-speed maps for the entire country, county by county, along with information about the location of the electricity network at both national and county levels. Distributed to all local authorities, the wind atlas was designed to help identify areas most suitable for wind energy developments. These maps are speculative. They represent the potential energy in kilowatt-hours that could be harnessed from the wind over a year, leveling out the peaks and troughs. Unlike coal or oil deposits that have a relatively fixed geographic (and geologic) location and volume that can be represented on a map, wind cannot be geographically located. What can be located, though, are turbines capable of harnessing energy from the wind that flows over the blades. The wind atlas, then, maps not the wind but the energy potential of the wind in specific geographic locations over a period of time.[4] The function of the map is to assess and visualize the potential of Ireland's west coast as a new energy frontier, a familiar move in the long history of colonization and enclosure: the identification of blank territories and the mapping of resources that lie idle and are in need of "improvement." Thus, the very calculation that just 2 percent of wind energy could solve world energy problems is enough to excite and mobilize environmental groups, politicians, energy companies, turbine manufacturers, and energy-hungry sectors of the economy (Cocozza 2017).

Since the 1990s, Ireland has halved its carbon emissions from electricity generation, mostly through onshore wind. Despite these developments, Ireland is set to miss its EU emissions targets for 2020 by a wide margin, potentially resulting in EU fines of hundreds of millions of euros.[5] To meet these targets, the government aims to *scale up* wind energy generation. Supporting such a move is an active wind energy market in Ireland that has begun to attract bigger investors such as Mitsubishi, Statoil, and General Electric, as well as pension funds seeking low-risk, long-term investments (Brennan 2017). But as the state, private energy developers, investors, and large environmental NGOs get behind the scaling up of wind energy capacity, contradictions have begun to materialize in the form of local opposition to new wind farm developments. Often dismissed as NIMBY politics, these localized conflicts and disputes challenge the development of wind farms on many grounds, including disturbance to the landscape, devaluation of property, impacts from light and sound, and effects on bird populations. In response, the state has begun turning its attention to a new frontier of wind energy: offshore.

Harnessing the more constant and intense power of the ocean winds, offshore "wind parks" promise to be a game changer in the field of wind energy, removing existing obstacles in the way of wind farm expansion and opening the way for larger flows of capital investment.[6] Environmentalists, communities affected by potential wind farm developments, and policymakers are in agreement that offshore is where wind energy development needs to go. As one journalist put it: "Offshore, there would be only the gulls to offend, and the people who will live, in four-weekly shifts, on the new accommodation vessels that are being deployed to manage the farms' growing distance from shore" (Cocozza 2017). But is there more at stake?

The move offshore poses new challenges, beginning with exclusive rights over large areas of Ireland's marine territory. In the United Kingdom, the Hornsea Project is a planned offshore wind park that will span

480 square kilometers. If greater quantities of oceanic wind energy are to be harnessed, then there needs to be a market to consume it. This potentially means price subsidies and price guarantees for private energy providers that have long been part of the political economy of energy. Financial supports will mean nothing without an energy grid capable of managing the quantity *and* quality of wind energy generated by offshore wind parks. This marks the principal stumbling block for producers and proponents of wind energy: the elusiveness of wind, its tantalizing presence and absence that promise abundant power in some moments and nothing in others. How can such an erratic element become the constant work/energy driving the on-demand, global economy?[7] If the energy potential of the wind on the west coast of Ireland is to be realized, there must be a radical transformation in how the current electricity grid and, by extension, the users of electricity operate. At the heart of this "problem" is the aligning of unpredictable supply of energy with what has historically been a relatively fixed demand. To adapt the grid to the erratic qualities of wind will involve both an extension of the grid (incorporating diverse renewable sources of energy) and a more fundamental transformation in the "intelligence" of the grid to align supply and demand in real time.

WIND WAS ONCE AMONG the principal sources of energy for North Atlantic economies, although it was not referred to as such. In 1800, five thousand windmills operated in England alone. The energy generated by these windmills was used to transform raw materials that needed pounding, mauling, shredding, hacking. or mixing; there were paper mills, mustard mills, hemp mills, grain mills, snuff mills, cocoa mills, oil mills, chalk mills, paint mills, and sawmills. Wind was also crucial for powering colonial ventures and mercantile capitalism: not just manufacturing but mobility and transport, of slaves, raw materials, and processed commodities across the world. Eighteenth-century sailing ships were both sophisticated technologies for harnessing the wind and harbingers of the hierarchical organization of the factory.[8] It is no surprise, then, that in 1800 the focus of scientific inquiry relating to wind was on weather and navigation.[9] Storms and weather were a huge risk to trade and commerce, forcing boats to stay in port or throwing them off course. Wind was particularly a problem when coupled with other natural flows and rhythms—most major ports are tidal, being at the mouth of rivers, which meant that boats often had to wait for a full tide that coincided with an onshore wind. Coal-powered boats that began operating in the first half of the nineteenth century were not only bigger, faster, and able to move straighter but, more importantly, capable of moving *when needed*, revolutionizing global trade and logistics.

Coal (and the machines it powered) overcame the "limits" of wind (and other preindustrial sources of energy) by being both mobile and constant: William Blake's "dark satanic mills" were powered not by wind but by steam and thus able to operate twenty-four hours a day. As recent historical work has shown, the expansion of the coal-powered steam engine also did away with the need for artisanal labor—the worker who was more independent, less receptive to discipline, and unreliable and who could not be worked around the clock (Malm 2016). With coal, it was possible to generate work/energy anywhere, including the cities where labor was cheap and plentiful; coal was more consistent, seemingly abundant, and easier to discipline. Thus, coal-powered technologies displaced water- and wind-powered mills, as well as the diverse forms of work, culture, and collective life that were more aligned to seasonal, ecological, and climatic rhythms than the temporality of industrial production.

Economic arguments for the adoption of coal in early nineteenth-century Britain relied on the new science of thermodynamics, developed to understand heat and temperature and their relationship to work and power (Caffentzis 1973). Thermodynamics provided a technoscientific apparatus (a complex of material and technological applications and practices) that allowed the comparison of different activities, bodies, and machines in terms of efficiency, an inhuman property measured as units of work/energy required per unit of output. This new, general view of energy revolutionized the old mode of production that had assumed a fixed limit on the forms of energy that could generate work—human, animal, wind, and water. Turning to the seemingly limitless supply of coal (and capacity of associated machines) to perform work opened a new field of inquiry and innovation involving engineers and industrialists striving to improve the efficiency of coal-powered machines.

Framed in terms of efficiency gains, the big *E* energy narrative of coal (and later oil) obscured and repressed what Lohmann and Hildyard (2014) call small *e* energy narratives—that is, the specific arrangements of people, places, animals, elements (including wind), and biomass that were organized according to different social and ecological practices, values, and contexts. By the end of the nineteenth century, "the commensurability of heat, kinetic energy, electricity, magnetism, and chemical energy had been cemented not just in the discourse of physicists, but also in the materiality of everyday life, in the conversion engines, factory floors, wires, pipes, travels and relationships linking goods, metals, and peoples and their territories, routines and governments across the world" (Lohmann 2016, 46). It was through this material and discursive assemblage that the category of energy was established, becoming part of the more general currency of "nature" during the nineteenth century. These developments were not just manufactured by capitalist interests; they relied on a host of technoscientific interventions, novel fields of expertise, technical practices, and infrastructures that both opened up a new frontier for work/energy *and* channeled it toward the political economic interests of industrial capital and *against* the interests of factory workers, craftspeople, urban dwellers, and people in the colonies, who were then forced to supply greater quantities of raw materials to feed the coal-powered factories.

The big E energy narratives that supported the emergence of coal power in the nineteenth century now operate at the nexus of energy and environment, making it appear inevitable and necessary that energy frontiers will need to be extended into the atmosphere; at stake are not just economic productivity and efficiency but the mitigation of climate change and, ultimately, the survival of the species. But the historic "limits" of wind as a source of energy have not gone away. Wind continues to be unpredictable and unruly. What makes it energetic also makes it resistant to stable energy production. The temporal and spatial uncertainty and specificity of wind remain the principal challenge for engineers and scientists tasked with harnessing the potential energy of wind and putting it to work for the good of present and future generations. If the nineteenth-century "miracle" of combustion involved the controlled *release* of energy from coal, then the twenty-first-century energy question hinges on the controlled *storage* of energy that flows all around us. And if the steam engine was the machine technology that enabled a coal-powered energy system, then it is the algorithms controlling the smart grid that will enable a wind-powered energy system.

At a European level, the development of clean energy is part of the elusive bid to transition toward a smart, innovative, and sustainable economy. As part of this transition, the EU envisages a seamless electricity grid stretching from the west coast of Ireland to the eastern borders of Romania. In 2009, the European Climate Foundation commissioned the Office for Metropolitan Architecture to create a graphic representation of what this future might look like. The resulting map describes European regions ("Eneropa") according to type of energy generation. In northwest Europe there is "Geothermalia"; "Solaria" stretches across the Mediterranean south; the UK and Ireland become the "Tidal States" and the "Isles of Wind." There is something folkloric about the renaming of geographical regions on the basis of the energies they possess and something highly ambitious about the prospect of these diverse energies flowing as clean, abundant electrical energy through a common, smart grid. The map recalls the journey of Nick Hunt to find the four winds whose names describe ancient routes and landscapes. Rather than return us to a time of awe and wonder, however, Eneropa represents a vision of the future where wind (along with the elements of water, earth, and fire) can be seamlessly enrolled within new energy systems such that those who use the energy produced will not notice the difference (Barber 2013).

To incorporate intermittent energy resources such as wind, electricity networks will have to become "smarter," with integrated communication systems, storage capacity, and real-time balancing between supply and demand. This has a ripple effect in terms of energy infrastructure as the technologies and networks required to cope with ambient energy are redesigned or overlaid with smart devices capable of sensing and predicting real-time shifts in the supply and demand of energy. Batteries have become a major topic of research and innovation, most notably for the promotion of electric cars but also for better managing and compensating the flows of energy in the grid.[10] The term *battery* usually refers to a device using chemicals (ordinarily lithium) but can mean any kind of power storage capacity. Norway has sought to brand itself as the "green battery" of Europe, able to absorb surplus energy when demand is low by pumping water into high mountain lakes, which can then be released when demand increases (Brown 2015).[11] Similarly, Compressed Air Energy Storage (CAES) is a new form of design engineering that uses geological and postindustrial landscape features, such as salt caverns or disused mines, to store energy.[12]

When we talk of a smart grid, we are referring to flows of data as much as we are flows of energy. The promise of aligning the erratic movements of the wind with the needs of energy consumers relies on the generation and sorting of real-time data and medium-term forecasting of meteorological conditions. The need for greater quantities and more accurate data to better predict and align demand and supply of energy foregrounds the role of data infrastructures, networks of sensors, and algorithms capable, for example, of forecasting wind speeds and plotting them against predicted energy demands. The impulse to create data-driven environments (5G, internet of things, automated transport/farming, and so on) requires the harnessing of more wind energy, which in turn depends on the creation of smart grids capable of conditioning unpredictable winds with the on-demand economy. The atmosphere is drawn into the cloud, a Möbius strip of wind/data (Gabrys 2014a).

Cymene Howe writes that wind is air in motion and thus, unlike the land, cannot be fenced in and owned. "Wind is, by definition, unenclosable," she writes; "the corpus of the wind, its scant materiality, its mercurial existence, make for a different sort of commons—one that is perhaps particularly resistant to true enclosure" (Howe 2016, 8). But just as the earth had to be mapped vertically in order to be mined, so now is the atmosphere being mapped and operationalized so that its powers can be harnessed. As Derek McCormack (2017) has written, this latter arrangement differs from the past insofar as the infrastructures involved are making new kinds of use of the movement and materiality of the elements in order to generate and capture different kinds of value. Though the wind itself may not be enclosed, the infrastructures currently being developed to compensate for its "mercurial existence" are performing their own form of enclosure through networked forms of intermediation and control.[13] Without the smart grid, wind remains a localized, limited source of energy whose potential can never be fully realized; it is the *infrastructuring* of wind that renders it as universal work/energy, something more than local and thus exchangeable and investable.

It becomes possible to see how the harnessing of wind energy today represents the extension, even culmination, of a dream born in the early nineteenth century with the beginning of coal-steam power and thermodynamics: the promise of an infinite source of work/energy that could replace (and thus discipline) human labor, as well as the messy, conflicted, limited sources of energy, originally water and animal but now coal, oil, and gas. The picture that begins to condense is one of serried ranks of offshore turbines connected to data centers and automated container ships—the putting to work of limitless atmospheric energies (McDermott Hughes 2017).

But the extractive assemblage of wind as energy describes a complex geography of places, materials, people, and technologies required to transform the specificities, uncertainties, and vagaries of wind into a constant source of energy capable of powering the economy. In this context, the development of wind energy does not just involve localized exclusions generated through the enclosure of terrestrial and/or maritime territories where wind farms are constructed but a more extensive "regime of exclusion," as Tania Li (2014) terms it, required to deliver on the promise of abundant work/energy.[14] The principal source of this is the reliance on the same forms of heavy resource extraction that are responsible for global warming in the first place. The long-term environmental impacts of the extraction and disposal of these rare-earth minerals for turbines and batteries cannot be separated from the exploitative forms of labor involved (Alonso et al. 2012).[15] Less easy to measure are the unknown effects of large-scale wind turbines on human and environmental health, from bird life to marine ecosystems where offshore wind farms are planned. Already the term *wind turbine syndrome* refers to a constellation of symptoms including nausea, vertigo, tinnitus, sleep disturbance, and headaches generated by sound effects and the flickering of light in ways that were previously not possible and may take time to develop (Ottinger 2013). And what if wind parks develop on a larger scale at sea or even, as has been floated, up above in the jet stream, "the largest, most concentrated renewable energy source on the planet" (Cocozza 2017)? Such developments could create unpredictable, cumulative effects on marine ecosystems or even climate, preventing the interaction of winds with vast oceanic and atmospheric systems (Boehlert and Gill 2010; Possner and Caldeira 2017). Makarieva, Gorshkov, and Li (2008) note that enough large wind turbines could impact the water cycle: "When the moisture-laden ocean-to-land winds are, on their way to land, impeded by windmills, this steals moisture from the continent and undermines the water cycle on land; the more so, the greater the extent of the anthropogenic consumption of wind power. *In its effect, the use of wind power is equivalent to deforestation*" (286; emphasis added).

To paraphrase Papadopoulos (this volume), full scale renewable energy is an absolute necessity and simultaneously it is as a full-scale process impossible. Rather than being seamless and immaterial, such arrangements generate effects that may not become visible until it is too late. Just as the exploitation of fossil fuels relied on untold decomposing bodies, processed by the actions of microorganisms and geological forces (Huber 2013), not to mention the miseries and disruptions wrought by mining, drilling, and refining to extract this stored carbon, so too the development of (universal) wind energy has its invisible (and unknown) consequences that do not show up in carbon calculations.[16] For wind to be scaled to the global grid, "it will come, just as coal and then oil did, 'with its world,' not as a 'green,' innocent, sustainable, resource" (Stengers, this volume).

Tilting at windmills is an idiom in the English language that has its origins in Cervantes's picaresque novel, *Don Quixote*. In one of the chapters, the eponymous hero approaches some windmills. Believing them to be giants, he rides into the middle of them with his lance thrust out to attack. He is flung from his horse and nearly dies. Of all the misadventures in the book, it is this one that has retained its place in popular parlance, becoming shorthand for attacking imaginary enemies, or enemies that are incorrectly perceived as such. It is also used to describe an unfounded, romantic, or vain attempt to oppose adversaries, real or imagined, to see things that are not there, to invest qualities in things. In popular interpretations, then, *tilting at windmills* is usually used to denigrate a position or dismiss someone's perspective. Following Isabelle Stengers, a different reading of such "idiotic stories" pays attention to the way they slow down dominant accounts and framings, resisting "the consensual way in which the situation is presented and in which emergencies mobilize thought or action. This is not because the presentation would be false or because emergencies are believed to be lies, but because 'there is something more important'" (Stengers 2005, 185).

This call to "slow reasoning down," to "tilt at windmills," takes on a near-literal meaning today as opposition to wind turbines and allied energy infrastructures poses a challenge to the common-sense, urgent calls for action on climate change through large-scale energy transition. As in other parts of the world where potential wind energy is located, it is often peripheral populations (geographically and politically, if not socially and economically) who are most affected by these developments and who stand to benefit the least (Boyer 2011; Dunlap 2018; Franquesa and Bartolome 2018). Cymene Howe (2014) uses such a conflict on the Isthmus of Tehuantepec in Oaxaca, Mexico, to draw attention to the biopolitical dimension of large-scale wind energy development: the moral, and calculable, justification for large-scale wind energy developments (i.e., planetary sustainability, through reduced carbon emissions) versus local forms of life that opponents argue are directly threatened by such developments. In terms of the latter, opposition to wind energy developments is framed not simply in terms of aesthetic concerns or human health but as a disturbance to more complex "ecologies of emplacement." Such concerns are dismissed by developers and government officials as "igno-rance" (of climate change) or the residue of an Indigenous culture, result-ing in a contest over distinct ways of imagining and articulating what Howe (2014) describes as "anthropocenic ecoauthority"—about what counts, or should count, in making viable futures.[17] The "anti-eolic" po-sitions articulated by some local communities in Oaxaca that are about "protecting lifeways and ecological spaces" (2014) challenge whether large-scale wind energy projects should take place at all and question what other forms of life may be displaced, or sacrificed, in their name.

The wind sculptures of the Dutch artist Theo Jansen are removed from territorial struggles against neocolonial extraction. But the exper-imental practices that inform his work allow us to envisage other ways of working with wind and gesture toward a decolonizing of matter in the register of small *e* energy narratives. In 1990, Jansen wrote a col-umn in a Dutch national newspaper warning that sea-level rises caused by global warming might reflood Holland (Frazier 2011). As a solution, he proposed building animals capable of gathering and depositing sand on the dunes.[18] These animals would be propelled by the wind, gathering sand as they walked. The project, which was intended to last only a year, has now occupied Jansen for the past thirty years, giving rise to the fan-tastical "Strandbeests," or "beach animals." The Strandbeests are made out of the discarded tubular plastic casing used to cover electrical wiring in Holland since 1947.[19] Along with rising sea levels, this plastic casing is a monstrous byproduct of petroculture: nonbiodegradable plastics made from refined petroleum that will be part of our climate-changed planet for centuries to come (LeMenager 2013). Jansen discovered the many possibilities for this plastic casing as a child, when he used it as a blow-pipe to fire paper darts with messages on them. Now this plastic tubing has become the "protein" of the Strandbeests. As the Strandbeests have evolved, so too have the uses and combinations of the plastic tubing—to make muscles, skin, and bones.

There is early, fuzzy documentation of Jansen trying out the first Strandbeest, so-called Currens Vulgaris, in 1991 (see figure 7.1). The tubes are taped together, and in the footage we see Jansen supporting the an-imal because it cannot stand up independently in the wind. The move-ments of the sculpture are similar to those of a foal or calf attempting to walk for the first time, jittery and vulnerable. As outlined on Jansen's website (www.strandbeest.com), the Strandbeests have now experienced seven "evolutionary periods." Combining experiments on Scheveningen

FIGURE 7.1: Currens Vulgaris—1991. Courtesy of Theo Jansen.

Beach with computer modeling, Jansen has been able to arrive at the optimum proportion and length of the tubes that run from the horizontal spine to the shoe—what he calls the DNA of the basic Strandbeest. The current generation of Strandbeests possesses unexpected degrees of complexity: one beast can detect an incoming tide, turn, and retreat to higher ground; another plows an anchor into the ground in response to mounting winds. Mechanical nerves trigger reflexes that border on thought.

To survive, the Strandbeests must be able to cope with and respond to the wind. "The trick is to get that untamed wind under control and use it to move the animal," Jansen says (2008), and therefore the animal requires muscles (in the form of one plastic tube inside another) that can flex and lengthen. The Strandbeests have now evolved to such a degree that they have stomachs consisting of recycled plastic bottles containing air. Wings at the front of the animal (something like a mouth?) flap in the breeze, driving the pumps to fill up with a supply of potential wind over a period of several hours. The animals need to store only enough wind to survive, which means escaping the incoming tide when the wind has dropped.

It is not possible to understand the lifelike quality of the Strandbeests without seeing them in motion, propelled by the wind. On wet days when the wind does not blow and the Strandbeests are motionless, waiting for the weather to break and a breeze to pick up, they look like

FIGURE 7.2: Animus Omnia—2018. Courtesy of Theo Jansen.

playground toys or unremarkable public art. But when the wind blows, they come to life. This "aliveness" is not just a figure of speech. The combination of the wind and the particular arrangement of stiff plastic tubing produces movement that evokes "a shiver-inducing air of autonomy" (Frazier 2011). Jansen describes his work as part engineering and part sculpture. "I try to make new forms of life which live on beaches," he says, "and they don't have to eat because they get their energy from the wind" (Fairs 2014). Jansen contrasts the Strandbeests to an empty plastic bag that one might see blowing along the street. The bag is moved by the whim of the wind, but it does not move itself. The Strandbeests are also moved by the wind, but not without some design and intention. Jansen clearly has intentions for the sculptures, but there are limits on what he can do that are imposed by the form and function of the plastic tubing in relation to the wind. The Strandbeests are not solely the product of Jansen's imagination but a collaborative achievement resulting from interaction and negotiation with material and immaterial elements.[20] This way of working with the wind (and plastic) is closer to the kinds of small *e* energy narratives identified in the opening epigraph of this chapter. The wind is not work/energy but an energetic force that is brought into specific arrangement with other elements, creating specific effects that are not entirely replicable.

Over time, Jansen's work has become more oriented toward the Strand-beests' survival rather than the original goal of protecting the Dutch coastline from rising sea levels. The current refinements he is working on are designed, first and foremost, to help the Strandbeests "live their own lives," as he puts it. Connected to his art practice, an ethos of generosity is at work here. Jansen repeatedly talks about the Strandbeests as creatures with their own trajectory, so much so that they have now begun to evolve "behind his back." Jansen is referring here to the proliferation of Strandbeests around the world, which is largely due to his commitment to sharing pictures, descriptions, diagrams, and instructions, including the Strandbeests' all-important "DNA code," through his website. Thousands of people are now building Strandbeests, including smaller, 3D-printed versions, with many of them sharing their discoveries in turn. This clearly excites Jansen, who envisages Strandbeests spreading all over the world, still powered by the movement of air but not as reliant on coastal wind conditions or his limited understanding of their lifeways. "And we all think that we are doing it but, in fact, the Strandbeests hypnotise people to do this" (quoted in Fairs 2014). The Strandbeests are able to travel beyond the workshop and beach at Scheveningen not by "scaling up" but by translating into other contexts, remaining open to new forms of engagement and use (Tsing 2012).

In his essay "The Stories Things Tell and Why They Tell Them," Michael Taussig (2012) reflects on how working with materials can become a kind of aesthetic practice of reenchantment. He is referring specifically to extreme forms of labor that operate at the "intimate level of interaction with things"—of bodies colluding with materials in ways that redistribute the locus of agency and life in the same way that a shaman might instill life into inert objects. The idea of belonging as an intimate form of engagement (rather than property right or identity) between individuals and the materials they work with can help explain why Jansen has devoted three decades to experimenting with plastic piping and the way it interacts with the wind. As Jansen himself admits, if he were an engineer faced with the task of creating wind-propelled sculptures for moving sand, he would have finished years ago. The difference is that he would have needed to rely on many additional materials, including batteries. His concern, and the ethos that informs it, are focused on the more immediate limits and possibilities of the materials (and atmospheric energies) he is engaged with, rather than the more abstract goal of climate adaptation. It is through these intimate interactions that *his* goals have been generously disrupted. "Searching and fooling around is a long way of going about things; your destination has yet to be decided," Jansen (2008) says. "You will probably never arrive at a destination in the accepted sense of the word, but you are very likely to call in at places no one has ever been before."

As a speculative practice, this opens a whole new energy ecology—employing the detritus generated by the mixture of petrochemicals and large-scale energy systems with small *e* wind energy to create a new form of inorganic life. The question is no longer how we stop rising sea levels, but how we can live differently. As Jansen himself explains, it was through his ongoing interaction with the Strandbeests that he came to recognize them as having their own trajectories, as having qualities and interactions of their own that shaped their evolution, often beyond his control. By demonstrating a different way of working with material (plastic) and ambient elements (wind) that decenters the human "creator," the Strandbeests foreground the pleasure, frustration, and even betrayal that accompany making life with others, suggesting an ecological ethic that falls far short of the grand hopes of green modernization.

Maybe there are clues here: for returning to wind as unpredictable, erratic, and locally specific, not as something to overcome but as something to adjust to and experiment with. The difference here is that wind is not assumed to exist as potential work/energy but as something multiple and beyond anthropocentric interest, as something that might or might not collaborate in our projects. To reenchant wind means to work with these qualities not as obstacles in the way of the fantasy of infinite, green energy but as a moment to "tilt" at the insatiable search for new sources of work/energy. This also means reframing the ethics and politics of wind energy not just from the point of view of local opposition to infrastructure projects but as a challenge to the big *E* energy narratives and the anthropocentrism they rest on.[21] In the face of compelling new green energy narratives, driven by fears of climate change and pulled by promises of abundant clean energy, this is no easy task; knowledge of existing or potentially negative impacts may not be enough to disrupt wind's affective hold. Instead, what is required are more hopeful "political ecologies of the precarious," radical more-than-human collaborations capable of disrupting the damaging but powerful hold of "salvation" narratives (Weston 2012). It is in the reactivation of the latter, in the form of large-scale wind energy developments, that Jansen's Strandbeests provides some speculative antidote—not in seeking to develop a clean substitute for fossil fuels, but in refiguring wind as something more specific (than work/energy) and lively, as partial collaborator in more-than-human experiments that might only be "good enough" and always under revision.

NOTES

1. I am drawing on a broad body of work within geography, anthropology, and science and technology studies that seeks to undo the assumed "naturalness" of the objects of modern, technoscientific knowledge, particularly the "resource-ness" of extrahuman nature (Bakker and Bridge 2006; Li 2014; Moore 2017; Plumwood 1993). As scholars from these fields have argued in different ways, the bifurcation of "nature" (domain of science) and "society" (domain of politics) lies at the heart of the "modern constitution" and the extension of the "biopolitics of improvement" (de la Cadena 2015, 2017).

2. In this indifference, there are echoes of Anna Tsing's (2015) "plantation logics," the violent segregating and combining of certain parcels of land, species of sugarcane, and Black populations from the social, cultural, and political ecologies of which they were a part, in order to "scale up" production of a single commodity, to extract, and to accumulate.

3. This early venture was commissioned by Bord na Mona, the state agency with responsibility for exploiting Ireland's peat bogs for energy generation. At the time, Bord na Mona's chief executive officer was Eddie O'Connor, an early pioneer of wind energy who subsequently left the state agency and now runs one of the world's biggest renewable energy companies, Mainstream Renewable Power.

4. Other visualizations of the wind do represent the flows of wind, such as www.windy.com.

5. Latest analysis by the SEAI shows Ireland will miss its total 2020 renewable energy target by at least 3 percent. The EU will impose a penalty of up to €120 million for every 1 percent the state falls below target.

6. The size of the turbines planned for offshore wind parks dwarfs what is currently allowed under planning legislation onshore. Brian Britton, chairman of the recently formed National Offshore Wind Association, has said that offshore energy could be the future equivalent of Ireland's "tech and pharma success story" (Rowe 2017).

7. Currently, the inconsistency of wind means that conventional sources of energy must still be kept in supply for when the wind stops blowing.

8. The sailing ship was also a site of insubordination, mutiny, and piracy (Linebaugh and Rediker 2000). Sailors knew they could navigate the boat without the captain, while the captain was lost without sailors. This adds an interesting prehistory to the relationship between capital, energy, and political agency to complement the work of Timothy Mitchell (2011), which examines the political forms

constituted through energy systems—namely, the role of carbon (coal, then oil) in shaping mass democratic movements in the United States and Europe.

9. What mattered at this point was not the speed of the wind (its energy) but the qualitative wind conditions—what the wind did to boats and the sea. Beaufort's scale was an attempt to standardize the effects of the wind on the sails and on the sea in order to better monitor and map its distributed effects.

10. In 2015, Tesla embarked on the construction of what has been described as the "world's biggest battery." The one-hundred-megawatt lithium-ion battery was commissioned by the transmission company for South Australia in response to a major blackout last year. South Australia is the most wind-dependent state in the country and thus requires a more flexible grid for when the wind does not blow.

11. Norway currently has more kilometers of pipes carrying water to its hydro-electricity plants than it has miles of road, so controlling the flow is the key (Brown 2015).

12. Project CAES Larne, in Northern Ireland, has been designated as a European Project of Common Interest and recommended for grant funding of up to €6.5million under a new European financing instrument called Connecting Europe. The project makes use of salt deposits on the Antrim coast that are over 1,400 meters below the ground.

13. One way of understanding the infrastructural transformations required to overcome the age-old "limitations" of winds' uncertainty, intermittency, and geographic specificity is in terms of recent theorizations of "platform" capitalism (Srnicek 2017). As with platforms in general, the "smart" energy grid of the future requires scale: the more sources of renewable energy and the more users it incorporates, the more efficiently it will function—washing machines operate when the wind blows, and mountain reservoirs fill when the population sleeps.

14. Raj Patel and Jason Moore's (2017) work on "cheapening" as a set of intersecting strategies for controlling the "web of life" is also instructive. This analytic allows for a world-ecological understanding of capitalist production, including the key role of technoscience in creating new frontiers of "nature-as-resource" or putting "nature to work."

15. In a report (EPA 2013), the US Environmental Protection Agency's Design for the Environment Program concluded that batteries using nickel and cobalt, like lithium-ion batteries, have the "highest potential for environmental impacts." It cited negative consequences such as mining, global warming, environmental pollution, and human health impacts.

16. As researchers in critical media studies have shown for some time, the operations of the "cloud" are anything but immaterial (Cubitt 2016; Hogan 2015).

17. This can be seen as part of a long history of resistance to "improvements." In the context of energy, for example, the Luddites were nineteenth-century textile workers and weavers who destroyed the new coal-powered machines. This "resistance was never against technology as such," as Bonneuil and Fressoz (2016, 261) remind us, "but against a particular technology and its ability to crush

others"—that is, the capacity of these technologies to make other worlds less possible, even dead.

18. It is perhaps no surprise that the Strandbeests were born in Holland, a country with a long tradition of wind technologies. The wealth of the country was founded in the seventeenth century on global trade that relied on the wind both for processing raw materials and for circulating them around the world. Most of the windmills, however, were used not to process raw materials but to repel the sea from the land, perhaps an early inspiration for Janssen's own Strandbeests.

19. It is estimated the Dutch have produced six million kilometers of this plastic tubing since then, much of it discarded (Jansen 2008).

20. As Howe writes in her essay "On Kinetic Commons," there is a need to release the concept of *the commons* from *terra*, a stable ground: "Rather than dominion and rights-to, this would be to think in terms of a commons as contact and interaction with" (Howe 2016, 1). *Commoning* in this sense is not so much about sharing a resource as sharing a concern, of constructing mutualistic collaborations with particular human and nonhuman others.

21. Melinda Cooper encapsulates this challenge: "It therefore becomes urgent to formulate a politics of ecological contestation *that is neither survivalist nor techno-utopian in its solutions*" (Cooper 2011, 50; emphasis added). The challenge of politicizing the environment around a recognition of nonhuman limits (and agencies) is not only about avoiding the trap of "austerity" but of appealing to people's affective desires and hopes for the future.

REFERENCES

Abrams, Meyer H. 1957. "The Correspondent Breeze: A Romantic Metaphor." *Kenyon Review* 19 (1): 113–30.

Alonso, Elisa, Andrew M. Sherman, Timothy J. Wallington, Mark P. Everson, Frank R. Field, Richard Roth, and Randolph E. Kirchain. 2012. "Evaluating Rare Earth Element Availability: A Case with Revolutionary Demand from Clean Technologies." *Environmental Science and Technology* 46 (6): 3406–14.

Bakker, Karen, and Gavin Bridge. 2006. "Material Worlds? Resource Geographies and the Matter of Nature.'" *Progress in Human Geography* 30 (1): 5–27.

Barber, Daniel A. 2013. "Hubbert's Peak, Eneropa, and the Visualization of Renewable Energy." *Places Journal*, May.

Boehlert, George W., and Andrew B. Gill. 2010. "Environmental and Ecological Effects of Ocean Renewable Energy Development: A Current Synthesis." *Oceanography* 23 (2): 68–81.

Bonneuil, Christophe, and Jean-Baptiste Fressoz. 2016. *The Shock of the Anthropocene: The Earth, History and Us*. London: Verso.

Boyer, Dominic. 2011. "Energopolitics and the Anthropology of Energy." *Anthropology News* 52 (5): 5–7.

Brennan, J. 2017. "Japanese Buy €300m Irish Wind Farm Portfolio." *The Irish Times*, July 31.

Bridge, Gavin. 2001. "Resource Triumphalism: Postindustrial Narratives of Primary Commodity Production." *Environment and Planning A: Economy and Space* 33 (12): 2149–73.

Brown, P. 2015. "Norway Cranks Up Plan to Be Europe's 'Green Battery.'" *Climate Home News*, July 27.

Caffentzis, George. 1973. "The Work/Energy Crisis and the Apocalypse." *Midnight Oil: Work, Energy, War* 1992:215–71.

Cocozza, P. 2017. "Wild Is the Wind: The Resource That Could Power the World." *Guardian*, October 15.

Cooper, Melinda E. 2011. *Life as Surplus: Biotechnology and Capitalism in the Neoliberal Era*. Seattle: University of Washington Press.

Cubitt, Sean. 2016. *Finite Media: Environmental Implications of Digital Technologies*. Durham, NC: Duke University Press.

De la Cadena, Marisol. 2015. *Earth Beings: Ecologies of Practice across Andean Worlds*. Durham, NC: Duke University Press.

De la Cadena, Marisol. 2017. "Uncommoning Nature." *E-flux Journal* 65 (May–August).

Dunlap, Alexander. 2018. "The 'Solution' Is Now the 'Problem': Wind Energy, Colonisation and the 'Genocide-Ecocide Nexus' in the Isthmus of Tehuantepec, Oaxaca." *International Journal of Human Rights* 22 (4): 550–73.

EPA. 2013. "Application of LifeCycle Assessment to Nanoscale Technology: Lithium-ion Batteries for Electric Vehicles." Design for the Environment Program, EPA's Office of Pollution Prevention and Toxics. April 24, EPA 744-R-12-001. https://archive.epa.gov/epa/sites/production/files/2014-01/documents/lithium_batteries_lca.pdf.

Fairs, Marcus. 2014. "'I Try to Make New Forms of Life,' Says Strandbeests Creator Theo Jansen." *Dezeen*, December 12. https://www.dezeen.com/2014/12/12/strandbeests-theo-jansen-interview-wind-powered-machines-new-species/.

Franquesa, Jaume, and Jaume Franquesa Bartolome. 2018. *Power Struggles: Dignity, Value, and the Renewable Energy Frontier in Spain*. Bloomington: Indiana University Press.

Frazier, Ian. 2011. "The March of the Strandbeests." *The New Yorker*. August 29. https://www.newyorker.com/magazine/2011/09/05/the-march-of-the-strandbeests.

Gabrys, Jennifer. 2014a. "A Cosmopolitics of Energy: Diverging Materialities and Hesitating Practices." *Environment and Planning A: Economy and Space* 46 (9): 2095–109.

Gabrys, Jennifer. 2014b. "Powering the Digital: From Energy Ecologies to Electronic Environmentalism." In *Media and the Ecological Crisis*, edited by Richard Maxwell, Jon Raundalen, Nina Lager Vestberg, 3–18. New York: Routledge.

Hogan, Mél. 2015. "Data Flows and Water Woes: The Utah Data Center." *Big Data and Society* 2 (2): 2053951715592429.

Howe, Cymene. 2014. "Anthropocenic Ecoauthority: The Winds of Oaxaca." *Anthropological Quarterly* 87, no. 2: 381–404.

Howe, Cymene. 2016. "On Kinetic Commons: Unthinking Terra." Paper presented at the annual meeting of the American Anthropological Association, Minneapolis, November 18.

Huber, Matthew T. 2013. *Lifeblood: Oil, Freedom, and the Forces of Capital.* Minneapolis: University of Minnesota Press.

Hunt, Nick. 2017. *Where the Wild Winds Are: Walking Europe's Winds from the Pennines to Provence.* London: Nicholas Brealey.

Jansen, Theo. 2008. "Strandbeests." *Architectural Design* 78 (4): 22–27.

Klein, Naomi. 2015. *This Changes Everything: Capitalism vs. the Climate.* New York: Simon and Schuster.

LeMenager, Stephanie. 2013. *Living Oil: Petroleum Culture in the American Century.* Oxford: Oxford University Press.

Li, Tania Murray. 2014. "What Is Land? Assembling a Resource for Global Investment." *Transactions of the Institute of British Geographers* 39 (4): 589–602.

Linebaugh, Peter, and Marcus Rediker. 2000.*The Many-Headed Hydra: Sailors, Slaves, Commoners, and the Hidden History of the Revolutionary Atlantic.* London: Verso.

Lohmann, Larry. 2016. "What Is the 'Green' in 'Green Growth'?" In *Green Growth: Ideology, Political Economy and the Alternatives,* ed. Gareth Dale, Manu V. Mathai, and Jose A. Puppim de Oliveira, 42–71. London: Zed Books.

Lohmann, Larry, and Nicholas Hildyard. 2014. *Energy, Work and Finance.* Dorset, UK: The Corner House.

Makarieva, Anastassia M., Victor G. Gorshkov, and Bai-Lian Li. 2008. "Energy Budget of the Biosphere and Civilization: Rethinking Environmental Security of Global Renewable and Non-renewable Resources." *Ecological Complexity* 5 (4): 281–88.

Malm, Andreas. 2016. *Fossil Capital: The Rise of Steam Power and the Roots of Global Warming.* London: Verso.

McCormack, Derek P. 2017. "Elemental Infrastructures for Atmospheric Media: On Stratospheric Variations, Value and the Commons." *Environment and Planning D: Society and Space* 35 (3): 418–37.

McDermott Hughes, David. 2017. *Energy without Conscience: Oil, Climate Change, and Complicity.* Durham, NC: Duke University Press.

Mitchell, Timothy. 2011. *Carbon Democracy: Political Power in the Age of Oil.* London: Verso.

Moore, Jason W. 2017. "The Capitalocene, Part I: On the Nature and Origins of Our Ecological Crisis." *Journal of Peasant Studies* 44 (3): 594–630.

Ottinger, Gwen. 2013. "The Winds of Change: Environmental Justice in Energy Transitions." *Science as Culture* 22 (2): 222–29.

Patel, Raj, and Jason W. Moore. 2017. *A History of the World in Seven Cheap Things:*

A Guide to Capitalism, Nature, and the Future of the Planet. Berkeley: University of California Press.

Peters, John Durham. 2012. *Speaking into the Air: A History of the Idea of Communication.* Chicago: University of Chicago Press.

Plumwood, Val. 1993. *Feminism and the Mastery of Nature.* London: Routledge.

Possner, Anna, and Ken Caldeira. 2017. "Geophysical Potential for Wind Energy over the Open Oceans." *Proceedings of the National Academy of Sciences* 114 (43): 11338–43.

Rowe, Simon. 2017. "Ireland's Race against Time to Avoid €360m EU Renewables Fine." *Irish Independent,* March 5. https://www.independent.ie/business /irelands-race-against-time-to-avoid-360m-eu-renewables-fine-35502638 .html.

Schwartzman, David. 2015. "From Climate Crisis to Solar Communism." *Jacobin,* December 1.

Srnicek, Nick. 2017. *Platform Capitalism.* New York: Wiley.

Stengers, Isabelle. 2005. "Introductory Notes on an Ecology of Practices." *Cultural Studies Review* 11 (1): 183–96.

Strandbeest. https://www.strandbeest.com/.

Taussig, Michael. 2012. "The Stories Things Tell and Why They Tell Them." *E-flux Journal* 36 (July).

Tsing, Anna Lowenhaupt. 2012. "On Nonscalability: The Living World Is Not Amenable to Precision-Nested Scales." *Common Knowledge* 18 (3): 505–24.

Tsing, Anna Lowenhaupt. 2015. *The Mushroom at the End of the World: On the Possibility of Life in Capitalist Ruins.* Princeton, NJ: Princeton University Press.

Weston, Kath. 2012. "Political Ecologies of the Precarious." *Anthropological Quarterly* 85, no. 2: 429–55.

Yusoff, Kathryn. 2018. *A Billion Black Anthropocenes or None.* Minneapolis: University of Minnesota Press.

8

CROWDING THE ELEMENTS

Cori Hayden

People in high density crowds appear to move with the
flow of the crowd, like particles in a liquid.
—Brian E. Moore, Saad Ali, Ramin Mehran,
and Mubarak Shah, "Visual Crowd Surveillance
through a Hydrodynamics Lens"

In the aggregate which constitutes a crowd there is in no
sort a summing up or an average struck between its elements. What really takes place is a combination followed
by the creation of new characteristics, just as in chemistry, certain elements, when brought into contact—bases
and acids, for example—combine to form a new body
possessing properties quite different from those of the
bodies that have served to form it,
—Gustave Le Bon, *The Crowd: A Study of the Popular Mind*

What are crowds made of? From nineteenth-century French crowd theory to twenty-first-century hydrodynamic simulations of high-density
crowd flows funded by the US Department of Defense, it seems that
crowds have persisted in their power to dissolve and recompose the fundamental elements of liberal social theory and notions of democratic

publics. Neither "the individual" nor "society" has been a particularly useful or durable unit of analysis in crowd theory. People in crowds behave as if "particles" in a "thinking liquid" (Moore et al. 2011); crowds are associations of "heterogenous elements" which, "like bases and acids" in a chemical reaction, combine to form something entirely new (Le Bon [1895] 2009, 16). Crowds, as with *sociality* more broadly, are composed of ideas and gestures that are "magnetized" to each other by "imitation rays" (Tarde 1903, 69–70). Crowds can be precipitated from crowd "crystals," and sometimes they are even rivers, or waves, or fire (Canetti 1962, 73–85). Crowd theories have long been extraelemental, in the pointed sense that they vividly recompose social theory's vocabularies for parts, wholes, and the ties that bind.

This all sounds dreamy from the queerly elemental, science studies-ish standpoint of this volume and the work that animates it. Conditions were not always ripe for such an embrace of crowd theory. Tainted by the overt racism and elitism of some of its early practitioners, its vexing "antiliberal" commitment to theories of suggestion and imitation (Borch 2012), and its explicit mobilization by early and mid-twentieth-century fascist and authoritarian leaders (Mussolini among them), crowd theory was held at arm's length in much of mid- to late twentieth-century social theory, as if contaminated with the very forces it sought to describe.[1]

But crowd theory has been reactivated as a heterodox archive for theory and method with what feels like increasing urgency (see, for example, Borch 2012; Brighenti 2014; Chowdhury 2019; Cody 2015; Dean 2016; Kelty 2012; Laclau 2005; Mazzarella 2010; Schnapp and Tiews 2006). Strong empirical, analytic, and political demands animate this reactivation. From the tidbits sprinkled above, you can imagine how crowd theory's idioms might seem particularly well suited to feminist and science studies–inflected arguments that sociality is constitutively *more than* social (see, among many, Murphy 2017; Papadopoulos 2018; Puig de la Bellacasa 2017). In nearby neighborhoods of science studies, the late nineteenth-century work of Gabriel Tarde has been particularly generative in efforts to rethink the social as immanent and constituted by heterogeneous ingredients, rather than as a superorganismic thing (Barry and Thrift 2007; Candea 2010; Latour 2005). And not least, social media, big data, and internet platforms are generating crowds—of people-data, of algorithmic aggregations, of disorganized labor—in ways that seem both familiar and new (Irani, Kelty, and Seaver 2012).

In much of this literature, the return to crowds was at first celebratory, infused with a sense of possibility from the standpoint of left, small-*d* democratic, and antiracist politics, as in the Arab Spring and the Occupy movements, the ascent of Syriza in Greece and Podemos in Spain, and the power of the Black Lives Matter movement. These formations energized many with the promise of crowd potency (boosted, in turn, in and through social media) for revolutionary emergence (see Dean 2016, among others). But the authoritarian turn in Egypt after the much-celebrated events in Tahrir Square, the political events in the North Atlantic in 2016 and 2017 (including Brexit and Trump), and the continuing rise of the Far Right across Europe, South Asia, and South America have brought the illiberalism of crowd potency immediately "back" to the forefront, in both an epistemological and a political sense: epistemological, insofar as crowd theory has long suggested that the tenets of "liberal" social and political theory (including the sovereign individual and rational publics) have perhaps never quite held, "descriptively" speaking; and political, because, in their association with fascism and authoritarianism, crowds have long been associated with many of the things that twentieth-century political liberalism was supposed to have militated against and vanquished—unbridled racism and antidemocratic political formations, among them.[2]

Crowd theory matters today in part because it reminds "us" of something that should, in fact, require no reminder, given the constitutive place of racism and settler colonialism at the heart of some prominent bastions of "liberal democracy": perhaps illiberalism has lived at the heart of liberalism all along. The "return" of the crowd *as a matter of concern*, and the "surging energies, light and dark" (Mazzarella 2017, 2) with which it has been associated, are prompting a lot of reflection these days not just about crowds as sociological entities with particular characteristics but, more interestingly, about the vital and destructive energetics with which we associate them and about many of the deeper and ambivalent questions that crowd theory has long raised. What are the materials, substrates, media, and dynamics that bind us to each other, durably, or in punctuated moments of collective effervescence? What and who constitute us, singly and multiply? Who and what are poised to exert force in this world? What makes us vulnerable to and with each other? Do we even have the tools to apprehend such things today?

These are, of course, political questions, and I have a sense that they might flow downstream (that is, somewhat effortlessly) with a trenchant set of critiques of neoliberalism(s). The "return" of the crowd (not as a thing that ever went away, because it did not, but as a matter of concern and as a locus of theorization) is unfurling in part in the wake and in the midst of globalized neoliberal projects which have explicitly waged wars, in multiple forms, on collectiveness—including on the social itself, on labor, on communities and peoples who were already marginalized and are now under intensified threat. In the United States, these projects have taken form in part as a decades-long conservative, right-wing, and overtly racialized (racist) war on public spending and taxation (especially of the wealthy) as loci for collective solidarity and a minimal, but nonetheless essential, social safety net (see Cooper 2017; Hohle 2012; Thomas 2017). The exaltation of individual, market "freedoms" over any notion of freedom tethered to equality and justice, as Wendy Brown (2018, 2019) has characterized US-based neoliberal projects, or, as Margaret Thatcher framed it in the UK, the exaltation of market freedoms over society itself, is meant to leave us alone—not "overburdened" by the social or the state, and instead left to pursue our lives as individuals and families, in Thatcher's infamous Hayekian formulation (see Brown 2019; Cooper 2017).

Breakdown, atomization, dissolution: neoliberal politics in their many forms are elemental, in a deconstitutive and reconstitutive way. And their effects might reasonably lead us to wonder: Is attention to the crowd "back" as an antidote or response to neoliberalism's atomizations, reductions, and isolations? A swing of the pendulum? Perhaps. But here is a slightly different question: What if this thing called neoliberalism (and it is many things) and these wars on "society" (which are many things) do not just atomize us? What if they have crowded us, too?

In order to think with that provocation, we need to address a prior question: What *does* it mean to think with crowds? There is something molecular about them, it seems, something more than human; when crowds come together, the viscosity is high (Saldana 2007). Crowd theory has, for this reason, long been anxious theory. It has a mood, and a universe furnished oddly, and an underdetermined politics. It is certainly not one thing. But two preoccupations animating much of this work seem important to its current salience. The first is the way that it puts the figure of the atomized individual "under strain," in Andrea Brighenti's (2014) apt formulation. From the late nineteenth century forward, many crowd theorists started from the proposition that when we find ourselves in crowds, however they are constituted (as physical masses, as a national crowd at a distance, as virtual swarms, and now through and magnetized by social media), we lose ourselves—more specifically, our rationality, our individuality, our capacity for discernment, our boundaries. We are, crowd theory tells us, vulnerable to others: to suggestion, to the sway of magnetism, and to emotional contagion (see Orr 2006). The effects may be compelling and terrifying at once (Canetti 1962); they might form the very basis of sociality itself (Tarde 1903); or, conversely, they may threaten to dissolve or destroy society as "we" know it (Le Bon [1895] 2009). But in any case, the power of this vulnerability seems to lie, for many crowd theorists, in a kind of mysterious ineffability—crowds are another order of thing, an unfamiliar and hard-to-understand phenomenon or site of transformation. This point seeps into the second preoccupation that calls my attention right now. What some crowd theory voices, sometimes despite itself, is a powerful unknowability: a recognition, perhaps with humility, that there are things we may not know how to know.

An anxious, unknowable potency, the dissolution of the social, the rise of aggregations that seem not just human: is it any wonder that crowd theory has recently come calling?

BEFORE THEY RECOMPOSE, THEY MUST DISSOLVE

Twenty-first-century "elements thinking," as Stefan Helmreich usefully glosses it in his contribution to this volume, is a way to call forth and think with entanglements or with "molecular-molar meshwork" (Helmreich, this volume; see also Puig de la Bellacasa, this volume). It is a refusal of divisions between science and the social. It signals an openness to the chemopolitical (see Murphy, this volume). This orientation, Helmreich observes, stands in marked contrast to the kind of elements thinking that spiked in the late nineteenth-century natural and social sciences (e.g., Durkheim's sociology, or hydrodynamics), which sought to identify the basic, irreducible building blocks at the root of more complex entities. In other words, much work in the late nineteenth-century social and natural sciences aspired "to scale up—one might say, to *compound up*—from elementary to more complex processes and forms" (Helmreich, this volume). *That* elements thinking theorized a starting point in the smallest, irreducible unit—an elementary social form, an individual, a water molecule—which does not change as it is scaled up or multiplied to compose larger entities (a body of water, or a society) (Helmreich, this volume).

The weird wonderfulness of such late nineteenth-century crowd theorists as Gustave Le Bon and Gabriel Tarde is that more and less directly, they stood in opposition to this way of thinking about composite wholes and the elements that make them up. For all of the differences between Le Bon and Tarde in orientation and sophistication, they both offered a theory of crowds that often stood markedly in opposition to a liberal (in the epistemological sense) sociological vocabulary of what, who, and *how* we are, on our own and in relation to others. At the heart of crowd theories' refusals of late nineteenth-century elements thinking was the dissolution of the individual itself. Crowd theory *is* a theory of deindividuation (Borch and Knudsen 2013; Dean 2016).

Organized around and heavily influenced by a burgeoning body of work in psychology and medicine on hypnotism, late nineteenth-century French crowd theorists took seriously (and some were slightly terrified by) the notion that perhaps everyone—not just "hysterics" and the otherwise pathologized—could be susceptible or vulnerable to suggestion, emotional contagion, and the sway of imitation. With the elaboration of an experimentally observable understanding of the vulnerability of the individual as a sovereign site of rationality, will, and autonomy, the idioms of sleepwalking, contagion, and suggestion began to enliven theories of how all manner of things worked and worked *on* *us*—from market speculation (where a theory of crowds had in fact been articulated as early as the 1840s), to the experience of urban density, to the mesmerizing and zombifying effects of mass consumption, to political revolution and the rise of workers' demands (see Schnapp and Tiews 2006).

For Le Bon, nothing other than the sciences of hypnotism could explain how reasonable people, when brought together in certain conditions, might do and be (terrible) things they would never do or be on their own. His wildly popular, polemic, slightly hysterical work *The Crowd: A Study of the Popular Mind* ([1895] 2009) is an easy target in any attempt to understand crowd theory's "miserable" fate in the halls of twentieth-century academia, as Christian Borch (2012) so memorably puts it. Like the multitude of writers and scientists whose work he set out to synthesize (plagiarize?) and popularize (vulgarize?), Le Bon was preoccupied by the turmoil that had followed the French Revolution, by the short-lived socialist uprising of the 1871 Paris Commune, and by the ascent of the popular classes as a new kind of mass and force. With their demands for equality and for workers' rights, their ascent into governance, and their capacity to take the streets, crowds and their power seemed very hard to put back in the bottle ("There is no power, Divine or human, that can oblige a stream to flow back to its source" (Le Bon [1895] 2009, [7]). The new "ERA OF CROWDS," Le Bon practically shouted, amounted "to nothing less than a determination to utterly destroy society as it now exists" (6–7). His analytic efforts seemed guided by anxious resignation: better to understand them, so as "not to be too much governed by them" (9).

Key to understanding them was understanding how crowds cause individuals to lose their faculties of reasoning and self-control, precisely as "in the case of the hypnotised subject, for whom the conscious personality has entirely vanished; will and discernment are lost" (Le Bon [1895] 2009, 18).[3] A pungent late nineteenth-century evolutionary racism permeated it all: "by the mere fact that he forms part of an organized crowd, a man descends several rungs in the ladder of civilization. Isolated, he may be a cultivated individual; in a crowd, he is a barbarian—that is, a creature acting by instinct" (19).

The insults piled on. Crowds are feminized; "Latins" are the most feminized of all; workers making unreasonable demands are no different than the "Esquimaux incapable of reasoning" (Le Bon [1895] 2009, 24, 42). It is compulsive and compulsory for any of us writing about Le Bon and crowd theory now to speak the insults aloud, so as to distance ourselves very much from them (see Borch 2012; Dean 2016; Mazzarella 2010).

And yet, for all of that awfulness, Le Bon continues to compel, as part of a recent return to the broader field of crowd theory of which he is a part—Tarde, Durkheim, Freud, Canetti, and many others—and the very strong sense that there might be something there for us now, for good or for ill. Before news outlets started overlaying Donald Trump's shouty profile with Le Bon quotations (see Ryan 2016), "the Frenchman" had already made a perhaps more unexpected return. The political theorist Jodi Dean, fueled by the energies animating the Occupy movement in New York's Zucotti Park in 2011, found something "ingenious" in Le Bon's work, despite the fact that he was an unapologetic "racist" and a "plagiarist"—namely, his rendering of the crowd as a "'provisional being formed of heterogeneous elements'" (Dean 2016, 9). That heterogeneity, and the provisional being that it constitutes, are what she wants to understand, and harness, and direct into sustained revolutionary energy. This political energy is not and will not be made of individuals. She writes, "Against the presumption that the individual is the fundamental unit of politics, I focus on the crowd" (4). To that end, she shows us what Le Bon's theory of crowd transformation can sound like when recuperated and rewired from within, as a story *for* radical democratic mobilization against the elite: "the crowd is more than an aggregate of individuals. It is individuals changed through the torsion of their aggregation, the force aggregation exerts back on them to do together what is impossible alone" (9).

Exactly. With a difference. If Le Bon painted a picture of people in crowds as less than, or reduced—not to something elemental or irreducible, but rather infantilized and "suggestionized"—he also argued that this very process made crowds into something *more than*, and radically different from, the sum of their parts. Hence the analogy to chemical reactions with which I opened this chapter: "In the aggregate which constitutes a crowd there is in no sort a summing up or an average struck between its elements. What really takes place is a combination followed by the creation of new characteristics, just as in chemistry, certain elements, when brought into contact—bases and acids, for example— combine to form a new body possessing properties quite different from those of the bodies that have served to form it" (Le Bon [1895] 2009, 16). The antireductionism and deindividualizing effects of this move are crucial. Le Bon, oddly seeming to anticipate Deleuze's *Difference and Repetition*, says, "The individual in a crowd differs essentially from himself" (19).[4]

Indeed, the notions of imitation and suggestion fueling Le Bon's observation ran right through the work of most of the late nineteenth- and early twentieth-century French crowd theorists (for a careful and thorough history of this literature, see Borch 2012). Not all of this work pathologized the processes of deindividuation that came to live at the heart of crowd emergence. Gabriel Tarde, whose many strands of thinking have lately been the subject of a vigorous, multipronged resurrection, essentially called forth an alternate universe organized around the composite principle of imitation-suggestion (Barry and Thrift 2007; Candea 2010). He too insisted on dissolving the individual as the irreducible starting point for understanding social life. Ruth Leys (1993, 281) says of Tarde's deployment of imitation and suggestion, "By dissolving the boundaries between self and other, the theory of imitation-suggestion embodied a highly plastic notion of the human subject that radically called into question the unity and identity of the self. Put another way, it made the notion of individuality itself problematic." This analytic commitment did not confine itself to the realm of the human. Extending Spinoza's idea of the monad in his own *Monadologie*, Tarde made clear that while the emergent disciplinary sciences were becoming attached to their own "final elements," each of those reductions, or foundations, was in fact a fiction: "those final elements at which all sciences arrive, the social individual, the living cell, the chemical atom, were final only to the eyes of their particular science; even themselves are composites" (Vargas 2010, 208). Society as aggregate (a thing that is the sum of its individual parts) was one of the targets of Tarde's insistence on the nonreductive nature of all entities. Tarde explicitly took issue with Durkheim's notion of society as superorganism, as if it were an overarching, prior, objectified "container" composed of individual elements (Candea 2010).[5] Instead, he argued that the social is immanent—bottom-up, but without a solid bottom—emerging in minute, infinitesimal relations of association (see Vargas 2010, 208). Thus, the dynamics of suggestion-imitation and contagion were not the pathological attributes of a debased crowd; rather, these processes lay at the heart of sociality *itself* (see also Laclau 2005)—hence Tarde's famous and ever-intriguing declaration that "society is imitation and imitation is a kind of somnambulism" in *Laws of Imitation* (1903, 87). In Tarde's world, we start and end in relations of imitation, association, and ever-multiplying "difference" that just "keeps differencing" (Vargas 2010, 209). Thus, "a society is always in different degrees an association, and association is to sociality, to *imitativeness* so to speak, what organization is to vitality, or what molecular structure is to the elasticity of the ether" (Tarde 1903, 69–70). One might imagine that this domain-meshing idiom—combined with his not very liberal displacement of the individual as the locus of agency and sociality— could help explain why Tarde "lost" the battle with Durkheim over who would get to define a twentieth-century science of the social.

Of course, Tarde's complex and heterodox conceptual universe is precisely why he has been reclaimed in so many arenas of late; for example, Bruno Latour (2005) has declared him the true paterfamilias of science studies as actor-network theory.[6] But beyond Tarde and science studies, crowd theory seems powerful and necessary now in no small measure because of its interlinked commitments to antireductionism and to the dynamics of imitation, suggestion, and contagion, unmoored (sometimes) from their association with pathology. The decentering of the individual and the potency of imitation-suggestion recur in calls to dust off crowd theory—to reactivate it, and with it, perhaps, our own crowd energies as well. Jodi Dean substitutes the crowd for the individual as her starting place for political mobilizing and analysis. Andrea Brighenti (2014, 68) argues that "the reassuring image of the individual as a 'building block' entering various social compositions does not hold. Crowd states make individuals *invisible*." Christian Borch (2012), for his part, suspects that imitation and suggestion are actually central concerns to which sociologists should now be attending.

Crowd theory is an archive of conceptual work that does not hold the individual steady, isolated, and sovereign. Its epistemological antiliberalism or illiberalism was the problem. Its illiberalism beckons.

THAT WHICH EXPRESSES BUT DOES NOT EXPLAIN

If crowd theories have been theories of deindividuation, they have also been theories of nonexplanation, or of a certain indeterminate unknowability. There is something ineffable, diffuse, and resistant to reduction about the crowds of crowd theory. In this ineffability, we might identify a reactivation of a different kind of "elements thinking."

Le Bon, for a start, was slightly flummoxed at the impasse in understanding that crowds presented (to paraphrase: it is relatively easy to know that crowds work in this or that way, but it is so much harder to know why!). But Elias Canetti's *Crowds and Power* (1962) shows us how to dwell in this impasse. Passionately interested in the phenomenology of being within crowds, Canetti's work too runs roughshod over the fundamental elements of liberal social analysis, not least by attending to how crowds make vulnerable the presumed boundaries between individuals. But this book is very different from—and resolutely indifferent to—the work of earlier crowd theorists cleaving to imitation-suggestion. Canetti's idiom is all his own. *Crowds and Power* is populated by crowds that want nothing more than to expand, as if they are something alien and hungry—they want to "feed on anything shaped like a human being" (16). Indeed, crowds do and do not consist of men as humans. Inflation is a crowd phenomenon, made of money and people simultaneously—both exalted and depreciated in the idiom of "the million." "In an inflation, the unit of money suddenly loses its identity. The crowd it is part of starts growing, and, the larger it becomes, the smaller becomes the worth of each unit" (186). The results can be terrifying. "In its treatment of the Jews National Socialism repeated the process of inflation with great precision" (188).

Canetti's (1962, 75) mythopoetic catalog of crowd formations across millennia weaves through other "collective units which do not consist of men, but which are still felt to be crowds." In fact, he provides us here with an elemental crowd theory, in the earth, fire, water, and air sense. Perhaps it is now a crowd theory for environmental-political catastrophe. Canetti writes of fire as a crowd, absorbing, growing, encompassing all in its path; he writes of crowds standing tall and still, uniform, menacing and menaced, in ways that are alternately beautiful and threatening. Corn (not your usual Galenic or witchy element) planted in rows is a crowd, marked by sameness and uniformity and often met with death, mown down by blade. Forests are crowds, dense, tall, and unmovable, like an army. And though he calls these things (corn, forests, fire) crowd symbols, their crowdedness is not metaphorical. Fire's attributes are not *like* the crowd's attributes; they *are* the crowd's attributes. "Fire is the same wherever it breaks out: it spreads rapidly; it is contagious and insatiable; it can break out anywhere and with great suddenness; it is multiple; it is destructive; it has an enemy; it dies; it acts as though it were alive, and is so treated" (77).

Water, in the form of the sea, rivers, and waves, is a crowd too: "The sea is multiple, it moves, and it is dense and cohesive. Its multiplicity lies in its waves; they constitute it. They are innumerable; the sameness of their movement does not preclude difference of size. The dense coherence of waves is something which men in a crowd know well. It entails a yielding to others as though they were oneself, as though there were no strict division between oneself and them. . . . The specific nature of this coherence among men is unknown. The sea, while not explaining, expresses it" (Canetti 1962, 80). Canetti's poetics (for I do not know what else to call them) are remarkable. Apparently underwhelmed by the decades of prior work attempting to explain "the specific nature of this coherence among men," Canetti, with the sea, evokes it instead.

Crowds, with waves and as waves, in this way seem adjacent to explanation. Stefan Helmreich, again my fellow traveler here, suggested as much in the wake of the election of Donald Trump (Helmreich 2020; García Molina and Cossette 2016). In a 2016 interview in *Cultural Anthropology Fieldsights*, and subsequently in the essay, "Wave Theory~Social Theory" (2020) he reflected on the reverberations between his own work on hydrodynamics and the recurring, insistent invocation of "waves" to speak to the political moment ("'populist waves,' 'waves of nationalist sentiment,' 'a wave of economic angst,' 'a Catholic wave to White House win,' 'a wave of angry white voters,' 'waves of protest,' 'a wave of hate crimes'") (García Molina and Cossette 2016).

Helmreich offers an acute reading of what this recurring invocation of waves does: it expresses, rather than explains. The figure of the wave wells up "when structural, analytic, or causal accounts are . . . difficult to settle upon" (2020, 318). If their invocation could prompt us to ask questions about causality, Helmreich wonders whether we even have the tools with which to answer. "Are critical anthropology's listening instruments always the right ones? Even if we anthropologists and other social theorists listen, do we know what we are hearing?" (García Molina and Cossette 2016).

What resources for thinking, for hearing, *are* adequate to this moment? Crowds and waves are not the only idioms in circulation right now that evoke and think with that which is diffuse, powerful, immersive and that are not always or easily reduced to explanatory models, building blocks and their composite forms, or the language of structure, logics, or formations (Murphy 2017). For a start, I think of Jackie Orr's (2006) work with and on panic, suggestion, and mediation; Christina Sharpe's (2016) meditations on racism as atmosphere, as the weather that makes it hard for Black bodies to breathe; Joseph Masco's (2010) work on the strange weather of the US security state; and, of course, a vast catalog of post-Deleuzian affect theory, among many other touchpoints. Such interventions are ways of dwelling in the constitution and effects of illogic, violence, mana, vitality, life force, nonindividuated relations, collectivity, and intensities. They prompt us to think about how solidarities and affinities are constituted through such forces. They are ways to theorize the surrounds that constitute us, that compose us, and that can energize and overwhelm us.

Crowds and crowd theory are not "meta-" to this catalog of ways of thinking and engaging in the world. They are right in the midst. Perhaps they are in the air.

CLOUD CROWDS

I have suggested thus far that the crowds of crowd theory engage elemental thinking in two senses: first, in the way they presumably dissolve and radically reformulate elemental theories of individual-society; and second, in the way that they constitute crowds as more-than-human mediums, and as an almost atmospheric modulation.

These two points deliver us directly to a third, and by now achingly obvious, observation: it is nearly impossible to evoke crowds in these ways today without contending with social media. Facebook, Twitter, WhatsApp, and other social media platforms are among the most potent activating mediums for crowds today, from the ways they help constitute masses on the street, to their production of algorithmically aggregated swarms of similarity, to their tremendous efficacy in enabling the "contagious" spread of highly charged affect. Social media crowd us right up.[7]

In this sense, we might say that "the cloud"—that multiply obfuscating term that points to the tangled infrastructures of data, the internet, and social media (Hu 2015)—is one of the many places where we find crowds today, in all of their potency and in their many forms, including and especially in the form of anxieties about them.

There is something elemental about this point too. In *The Marvelous Clouds*, media theorist John Durham Peters (2015) helped propel the resurgence of contemporary elements thinking by routing the "new" in new media through something both familiar and strange. He argues that new media, digital media, and social media (same, not the same) are not primarily sources of information and meaning; rather, they must be thought elementally—that is, environmentally and infrastructurally, as if they are habitat (4).[8] Peters brings earth, soil, fire, water, ozone, and clouds to the core of his engagement with digital media precisely because these elements evoke but do not explain (in his words, they "have meaning but do not speak" [3]). Thinking media as elemental, he argues, recasts the problems that we confront in and through them. "So-called new media do not take us into uncharted waters: they revive the most basic problems of conjoined living in complex societies and cast the oldest troubles into relief" (4). Peters names these oldest troubles "civilizational." I might say, more specifically, that the troubles we have made through social media (*What has Facebook wrought?*) are troubles we know from late nineteenth- and early twentieth-century crowd theorists, and thus they are elemental troubles of a more recent vintage.

In fact, as matters of concern, crowds and social media seem in some respects to be one and the same. The catalog of resonances is vast and constantly expanding, as I have argued elsewhere (Hayden 2021). Algorithmic filter bubbles immerse us in spaces of self-reinforcing and ever-amplifying similitude, producing quasi-Tardean socialities constituted in imitation and similarity (Seaver 2012, 2021). Facebook, Twitter, and other platforms have become tremendously effective, often more-than-human, "suggestionizing" forces, as Le Bon described crowds; the specter of the mob (even if the mob is the state) is never far from the surface (I am thinking particularly of the 2019 *New York Times* investigative report, "A Genocide Incited on Facebook," on the Myanmar military's use of social media to cultivate anti-Rohingya sentiment [Mozur 2019]). Forms of data generation that feed on that which "looks like a human"—bots, markets in Twitter followers, click farms, "fake" accounts that might be automated and/or human-made—are central to how Facebook and Twitter elicit data crowds. Even as we are microtargeted as consumers and political animals, we are not thereby "isolated": we are swimming in crowds, having been deindividuated and algorithmically recomposed within "similarity spaces" (Seaver 2021). Social media's ability to fuel and enable an almost uncontrollably rapid transit of affect, "violent antagonisms," fakeness, and "irrationality" (Phillips 2018) *is* the problem named by Le Bon's concerns with crowds as mediums, magnetized by the force of imitation-contagion. As Whitney Phillips (2018) says of metric-driven online news in the age of social media, "things traveling too far, too fast, with too much emotional urgency, is exactly the point."

As with Canetti's fire as crowd, the resonances between crowd theorists' analytics, openings, observations, and anxieties and what we experience in social media are not "mere" similarities. Perhaps social media should be called crowd media, because the imagination of (more than) sociality on which these companies bank is the crowd (Hayden 2021). Facebook's business model, after all, is based explicitly on the monetization of emotional contagion. The platform, as with so many of its counterparts today, sells attention and hence advertising by multiplying clicks and shares, which in turn multiply more rapidly the more intense the emotional affect involved (see Baldwin 2019; Lanchester 2017; Phillips 2018; Vaidhyanathan 2018). Not surprisingly, then, Facebook has routinely been used as ready-made experimental terrain for psychologists trying to understand how emotional contagion works (Kramer, Guillory, and Hancock 2014).

Facebook *is* the fire; it is also arguably the smoke. Either way, among the things that many of us suspect are burning are the foundational underpinnings of liberal democracy and "liberalism" itself, flawed as such formations have always been, and exceedingly so in their neoliberal extremes. We see this in part in an increasing (and disconcertingly LeBon-ian) alarm that social media's particularly crowd-y modality of "connection" through "contagion" (Hayden 2021) may be threatening the constituent elements of society itself. Critics of Facebook, including but certainly not limited to a chorus of its former employees, have become vocal about the ways that these platforms and the forces they fuel and feed upon might well be unraveling "the social fabric," and with it, the substrate of rationality on which political equality and democracy are based (Wang 2017). Media and internet scholar Siva Vaidhyanathan (2018) offers an adjacent critique in *Antisocial Media: How Facebook Disconnects Us and Undermines Democracy*; so does philosopher Byung-Chul Han (2017) in his book *In the Swarm*, in which he refers to the hot mess of atomized, lonely outrage and immediacy that has come to stand for "authentic" communication on social media as "the shitstorm"; and so do countless politicians, activists, writers, and many others, myself included, on a regular, profane basis. At the heart of many of these critiques, as with critiques of neoliberalism, is the persistent sense that society, and sociality, have been rent asunder; that the fundamentals of a social contract that might recognize and demand equality, not to mention collective responsibility for one another, are being torn up; that we are left with something elemental, or stripped bare—base "shittiness," perhaps; atomization; only isolated individuals and (maybe) our families.

We are, in these terms, currently experiencing an elemental (i.e., political) crisis. But what if, as crowd theory has long insisted, this base pair—individual and society—is not quite enough to name the troubles and the energies with which we are contending? What if these atomizations *crowd*, too? After all, at least in the United States, these ongoing atomizations and the racialized political economic formations that have fueled them have also reconstituted or even recharged a host of not new and not arbitrary illiberal solidarities, patriarchal whiteness prominent among them (Anderson 2016; Cooper 2017). The January 6 riot at the Washington, DC, Capitol, and the deadly theatricality of the white supremacists who marched in Charlottesville, Virginia, with their tiki torches, are of course merely among the more spectacular examples.

Thinking about social media almost symptomatically, as crowd potency and as crowd threat, certainly churns up some of the biggest political and social questions confronting us today, including the compositional questions I broached at the outset of this chapter: What *are* the materials, substrates, connecting forces, vulnerabilities, and dynamics that bind us to others? What is it that threatens or destabilizes such binding forces? These questions cannot be treated as if they are purely psychological phenomena, as Le Bon might have had it. They are political and political-economic questions; they are normative, descriptive, and conceptual questions; they are invitations for struggle. They are also historical—which is to say, specific. And, insofar as they are atmospheric, they are also (infra)structural. After all, if the cloud (as internet, as social media) is "elemental" in John Durham Peters's assessment, it is also, as Tung-Hui Hu (2015, 147) argues for the deregulated United States, a "metaphor for private ownership." That we should be so intensively and often pleasurably crowded by and through platforms that are, at their core, vehicles for private capital accumulation is crucial to, but not a reductive explanation for, our current compositional troubles.

Why, then, have crowd theory's crowdings come calling? Perhaps because crowd theory itself recomposes the elements of social action and analysis; perhaps because, as an archive of an "illiberal" twentieth-century social theory, crowd theory raises unresolved questions about the not so submerged "undersides" of political liberalism itself; perhaps because its underdetermination is confusing and unmooring, befitting a moment when new composites ("illiberal democracy" among them) are scrambling and rearranging the twentieth century's terms of political orientation; perhaps because the crowd, in fact, never went away as both an excessive and a necessary form of mass politics, even if the postwar North Atlantic world told itself a different story (Chowdhury 2019). Perhaps because, as Canetti noted, there is something very powerful about finding, and losing, oneself in a crowd; the question, as ever, is what we do with and through that potency.

NOTES

1. William Mazzarella (2010) notes the "contempt" with which crowds and their theories have been held in social theory, just as Christian Borch (2012) observes that crowd theory was banished to the "margins of respectable sociological theory" for a large part of the twentieth century.

2. *Illiberalism* is a complicated, troubling, and troubled term. I am invoking it here in a kind of composite sense. If crowd theory is arguably "illiberal" in the sense that it destabilizes the foundational figure of sovereign individuals and rational publics, it can also remind us that these liberal elements should not and cannot serve as the yardstick against which all other political forma-

tions are measured (for arguments that mass politics in India and Bangladesh are importantly and not pejoratively illiberal, see Chowdhury 2019; Cody 2015). It is in these senses that Christian Borch calls crowd theory antiliberal. But the term *illiberal* also has a renewed and specific life in normative political science and punditry, as, for example, in the form of a new composite, *illiberal democracy*. The term first surfaced as an accusation leveled in the late 1990s at "emerging" democracies which were considered not quite good (liberal) enough and which "still" bore the marks of preceding authoritarian regimes (Plattner 2019). That accusation has now come to be embraced by some of its targets, including right-wing figures such as Hungarian president Viktor Orban, who sees "illiberal democracy" as a pretty good description of his vision for an anti-immigration, pro-patriarchal-family "Christian democracy" that is not tainted by "Western" values (Plattner 2019). Alongside this trajectory we could point to what Papadopoulos, Stephenson, and Tsianos (2008) call postliberal projects, in which ostensibly liberal states make decidedly illiberal moves—revoking citizenship for targeted groups, suppressing the Black vote—in the name of protecting particular, narrowly conceived "freedoms."

3. The individual who finds himself in a "psychological crowd" is "no longer conscious of his acts"; "under the influence of a suggestion," he shows "irresistible impetuosity"—an irresistibility made all the stronger because of the multiplying, "reciprocal force" of a crowd whose members, through "suggestion and contagion," are all directed toward the same idea ([1895] 2009, 18–19). In a crowd, subjects lose their ability to deliberate and *then* act: it is all impulse, as the same idea immediately fuels the same action, accomplishing something that no people in their right mind would do outside of a crowd.

4. But the anticipation is actually not that odd. In *Difference and Repetition*, Deleuze drew directly on Gabriel Tarde, Le Bon's contemporary and fellow traveler in the elaboration of theories of imitation and suggestion, as a nonreductive way in to his reworking of sameness and difference.

5. Durkheim was a target, but it seems prudent to resist the polemical move to flatten him out in the rush to rescue Tarde from "obscurity"; see, for example, Mazzarella's (2017) beautiful reanimation of Durkheim in *The Mana of Mass Society*.

6. Anyone familiar with Bruno Latour's career-long war against the a priori idea of "society" as something that explains things happening at a "smaller," individual scale will understand the exuberance. Latour celebrates Tarde's monadology thus: "Tarde offers a very odd type of reductionism since the smallest entities are always richer in difference and complexity than their aggregates." "Because he does not stop at the border between physics, biology and sociology . . . , he does not believe in explaining the lower levels by the higher levels" (2005, 2).

7. Thanks to María Puig de la Bellacasa (personal communication with the author, November 8, 2019).

8. Peters (2015, 4) writes, "Digital devices invite us to think of media as environmental, as part of the habitat, and not just as semiotic inputs into people's

heads." This move cuts both ways: he also argues that the environment is a medium, constituted by human meaning-making practices but not wholly determined by them.

REFERENCES

Anderson, Carol. 2017. *White Rage: The Unspoken Truth of Our Racial Divide*. New York: Bloomsbury Press.

Baldwin, Tom. 2019. *Ctrl Alt Delete: How Politics and the Media Crashed Our Democracy*. London: Hurst.

Barry, Andrew, and Nigel Thrift. 2007. "Gabriel Tarde: Imitation, Invention and Economy." *Economy and Society* 36, no. 4: 509–25.

Borch, Christian. 2012. *The Politics of Crowds: An Alternative History of Sociology*. Cambridge: Cambridge University Press.

Borch, Christian, and Britta Timm Knudsen. 2013. "Postmodern Crowds: Re-Inventing Crowd Thinking." *Distinktion: Scandinavian Journal of Social Theory* 14 (2): 109–13.

Brighenti, Andrea. 2014. *The Ambiguous Multiplicities: Materials, Episteme and Politics of Cluttered Social Formations*. Camden, UK: Palgrave Pivot.

Brown, Wendy. 2018. "Neoliberalism's Frankenstein: Authoritarian Freedom in Twenty-First Century 'Democracies.'" *Critical Times* 1, no. 1 (April 1): 60–79.

Brown, Wendy. 2019. *In the Ruins of Neoliberalism: The Rise of Antidemocratic Politics in the West*. New York: Columbia University Press.

Candei, Matei, ed. 2010. *The Social after Gabriel Tarde: Debates and Assessments*. London: Routledge.

Canetti, Elias. 1962. *Crowds and Power*. New York: Farrar, Straus and Giroux.

Chowdhury, Nusrat. 2019. *Paradoxes of the Popular: Crowd Politics in Bangladesh*. Palo Alto, CA: Stanford University Press.

Cody, Francis. 2015. "Populist Publics: Print Capitalism and Crowd Violence beyond Liberal Frameworks." *Comparative Studies of South Asia, Africa and the Middle East* 35, no. 1: 50–65.

Cooper, Melinda. 2017. *Family Values: Between Neoliberalism and the New Social Conservatism*. New York: Zone.

Dean, Jodi. 2016. *Crowds and Party*. New York: Verso.

García Molina, Andrés, and Julien Cossette. 2016. "Election's Reverb: An Interview with Stefan Helmreich." Supplementals, *Fieldsights*, December 16. https://culanth.org/fieldsights/elections-reverb-an-interview-with-stefan-helmreich.

Han, Byung-Chul. 2017. *In the Swarm: Digital Prospects*. Translated by Erik Butler. Cambridge, MA: MIT Press.

Hayden, Cori. 2021. "From Connection to Contagion." Special issue, "The Anthropology of Data," guest edited by Antonia Walford, Nick Seaver, and Rachel Douglas-Jones. *Journal of the Royal Anthropological Institute* 27:95–107.

Helmreich, Stefan. 2020. "Wave Theory ~ Social Theory." *Public Culture* 32, no. 2: 287–326.

Hohle, Randolph. 2012. "The Color of Neoliberalism: The 'Modern Southern Businessman' and Postwar Alabama's Challenge to Racial Desegregation." *Sociological Forum* 27 (1): 142–62.

Hu, Tung-Hui. 2015. *A Prehistory of the Cloud*. Cambridge, MA: MIT Press.

Irani, Lilly, Christopher M. Kelty, and Nick Seaver, eds. 2012. "Crowds and Clouds." *Limn*, no. 2. https://limn.it/issues/crowds-and-clouds/.

Kelty, Christopher M. 2012. "Preface: Crowds and Clouds." *Limn*, no. 2. http://limn.it/preface-crowds-and-clouds/.

Kramer, Adam, Jamie E. Guillory, and Jeffrey T. Hancock. 2014. "Experimental Evidence of Massive-Scale Emotional Contagion through Social Networks." *Proceedings of the National Academy of Sciences* 111 (24): 8788–90.

Laclau, Ernesto. 2005. *On Populist Reason*. London: Verso.

Lanchester, John. 2017. "You Are the Product." *London Review of Books*, August 17, 3–10.

Latour, Bruno. 2005. *Reassembling the Social: An Introduction to Actor-Network-Theory*. Oxford: Oxford University Press.

Le Bon, Gustave. (1895) 2009. *The Crowd: A Study of the Popular Mind*. New York: Classic.

Leys, Ruth. 1993. "Mead's Voices: Imitation as Foundation, or, The Struggle Against Mimesis." *Critical Inquiry* 19 (2): 277–307.

Masco, Joseph. 2010. "Bad Weather: On Planetary Crisis." *Social Studies of Science* 40 (1): 7–40.

Mazzarella, William. 2010. "The Myth of the Multitude, or, Who's Afraid of the Crowd?" *Critical Inquiry* 36 (4): 697–727.

Mazzarella, William. 2017. *The Mana of Mass Society*. Chicago: University of Chicago Press.

Moore, Brian E., Saad Ali, Ramin Mehran, and Mubarak Shah. 2011. "Visual Crowd Surveillance through a Hydrodynamics Lens." *Communications of the ACM* 54 (12): 64.

Mozur, Paul. 2019. "A Genocide Incited on Facebook, with Posts from Myanmar's Military." *New York Times*, March 4.

Murphy, Michelle. 2017. "Alterlife and Decolonial Chemical Relations." *Cultural Anthropology* 32 (4): 494–503.

Orr, Jackie. 2006. *Panic Diaries: A Genealogy of Panic Disorder*. Durham, NC: Duke University Press.

Papadopoulos, Dimitris. 2018. *Experimental Practice: Technoscience, Alterontologies, and More-Than-Social Movements*. Durham, NC: Duke University Press.

Papadopoulos, Dimitris, Niamh Stephenson, and Vassilis Tsianos. 2008. *Escape Routes: Control and Subversion in the 21st Century*. London: Pluto.

Peters, John Durham. 2015. *The Marvelous Clouds: Toward a Philosophy of Elemental Media*. Chicago: University of Chicago Press.

Phillips, Whitney. 2018. *The Oxygen of Amplification: Better Practices for Reporting on Extremists, Antagonists, and Manipulators Online*. New York: Data and Society Research Institute. https://datasociety.net/output/oxygen-of -amplification/.

Plattner, Marc. 2019. "Illiberal Democracy and the Struggle on the Right." *Journal of Democracy* 30 (1): 5–19.

Puig de la Bellacasa, María. 2017. *Matters of Care: Speculative Ethics in More than Human Worlds*. Minneapolis: University of Minnesota Press.

Ryan, Verity. 2016. "How a 19th Century Frenchman Predicted the Rise of Donald Trump." *The Telegraph* (London), September 8. https://www.telegraph.co.uk /news/2016/09/08/how-a-19th-century-frenchman-predicted-the-rise-of -donald-trump/.

Saldana, Arun. 2007. *Psychedelic White: Goa Trance and the Viscosity of Race*. Minneapolis: University of Minnesota Press.

Schnapp, Jeffery T., and Matthew Tiews, eds. 2006. *Crowds*. Stanford, CA: Stanford University Press.

Seaver, Nick. 2012. "Algorithmic Recommendations and Synaptic Functions." *Limn*, no. 2. http://limn.it/algorithmic-recommendations-and-synaptic -functions/.

Seaver, Nick. 2021. "Everything Lies in a Space: Cultural Data and Spatial Reality." Special issue, "The Anthropology of Data," guest edited by Antonia Walford, Nick Seaver, and Rachel Douglas-Jones. *Journal of the Royal Anthropological Institute* 27:43–61.

Sharpe, Christina. 2016. *In the Wake: On Blackness and Being*. Durham, NC: Duke University Press.

Tarde, Gabriel. 1903. *Laws of Imitation*. Translated by Elsie Clews Parson. New York: Holt.

Thomas, Todne. 2017. "(Ir)Ration(Aliz)Ing African American Kinship: Neoliberal Moral Orders of Family and Race." Paper presented at the annual meeting of the American Anthropological Association, Washington, DC, December 2.

Vaidhyanathan, Siva. 2018. *Antisocial Media: How Facebook Disconnects Us and Undermines Democracy*. Oxford: Oxford University Press.

Vargas, Eduardo Viana. 2010. "Tarde on Drugs, or Measures against Suicide." In *The Social after Gabriel Tarde: Debates and Assessments*, ed. Matei Candea, 208–29. Abingdon, UK: Routledge.

Wang, Amy B. 2017. "Former Facebook VP Says Social Media Is Destroying Society with 'Dopamine-Driven Feedback Loops.'" *Washington Post*, December 12.

9

EMBRACING BREAKDOWN: SOIL ECOPOETHICS AND THE AMBIVALENCES OF REMEDIATION

María Puig de la Bellacasa

ORIGIN STORIES: FROM STARDUST TO LIFE

For my daughter's sixth birthday, we booked a children's party at the Space Centre, a museum in our city dedicated to space science and technologies. The gig included the viewing of an educational animated film called *We Are Stars*, shown in a planetarium-style immersion projection hall. The film was a cosmic Earth biopic, telling children and their adult companions a classic popular-science tale, that of the elementary origins of Life. Once upon a time, six entities known as the building blocks of life emerged from billions of years of cosmic fireworks and stardust mixings: oxygen, hydrogen, carbon, nitrogen, sulphur, and phosphorus. Fusions and bangs went on and on for so long a time that is impossible to comprehend; and even once Earth was formed as a planet, geochemical melt and shake followed ceaselessly, until a mysterious juncture allowed *geos* to engender *bios*. Something happened when elements organically compounded and recompounded into chemical substances, became involved with each other to generate forms teeming in the elemental sea (or soup), and came to be called life. The *We Are Stars* tale continued, with a commonplace plot and ending: the depiction of a tree of life shooting up into a myriad of straight branches and subbranches, holding living creatures of all shapes and sizes. At the top of one of the highest branches was the end product, the human figure. Us. Stars out of stars.

This story is commonplace to those acculturated in popular, modern Western cosmologies imbibed with aestheticized scientific themes and metaphors. From stardust to elements, from elements to compounds, to substances and life of all kinds, crowned by life with a capital *L*, animated by superior intelligence,—all emerge from primal stardust recompositions. This is a charismatic tale, telling how the multifarious forms of life that we know arose from what scientists now call biogeochemical processes. What's not to like? It is a wondrous, semidivine achievement, yet rationalized; a science-based, secular-materialist, origin story that nevertheless inspires mythical awe and reverence for the magical and improbable emergence and rise of life (a paradoxical occasion, both explainable and mysterious).

There is much to say about the narratives and imaginaries conveyed by this story. Imaginaries of cosmic wonder continue to generate joy and delight. Yet, something struck me as not particularly educational in today's planetary circumstances: not only the human exceptionalism that the story tends to perpetuate but, more subtly, the mobilization of the biogeochemical natural history of the elements as one of buildup toward achieved complexity. I felt unease at the absence of an indispensable dimension of the biogeochemical tale: that life on Earth as we know it is as much the creation, the buildup, of stuff as it is its elemental breakdown and recirculation. The imaginary that equates life to creation, growth, and attainment in this long-standing story missed integrating even today's most conservative conceptions of sustainability that translate into popularizations of ecological thinking (from living in harmony with natural cycles to ideas of circular economies): the breakdown and circulation of matter that rebalance generation, productivity, and excess in a finite Earth. Most important for children growing up in a postindustrial city of the Global North, *We Are Stars* does not tell how biogeochemical cycles that took eons to be established—the ecological choreographies forming Earth as we know it—slowly started to become affected by cultivation activities about ten thousand years ago and have become dangerously disrupted since the industrial and agricultural revolutions. Nor does it explain that the situation is dire, and the process seems unstoppable. Of course, as a parent, I feel one would have to find a way for new retellings that do not damp down cosmic wonder with paralyzing fear and humanist guilt but nurture a sense of involvement that transforms hope and joy into an everyday practice—a way that inspires curiosity about how, on Earth, we will continue to live together as a more-than-human community.

BREAKDOWN ECOPOETHICS

My sensitivity to the absence of this story line in the education of children, many of whom already know they will inherit a damaged and trashed planet, is made more acute by my interest in contemporary transformations in relation to soils in a time of environmental disruption. Through this research, I came to see soils as the embodiment of processes of creation-decreation. Soils are vital media for the growth of living earthy things that are, essentially, a result of the breakdown of mineral and organic matter. Thinking with soils, this chapter attempts to prolong motifs in the story of life as biogeochemical wonder. The first of these motifs is the sublime and trusting thought of the elemental affinity of humans and all other living beings and forms with the material existence of all that is and has been in earthy and cosmic realms—*we are stardust*.[1] The second is a fascination provoked by these captivating processes—more specifically, the awe at the buildup of matter from what we deem simple to what we deem complex, from emergence, to creation, to *production*. This chapter also, of course, seeks to disrupt, rather than prolong, the incongruous and smug thought of the human as pinnacle of cosmic achievement, which feeds on the way we tell these stories.

Soils creation as the ongoing result of biochemical processes of breakdown emphasizes decomposing, decompounding, and decreating and disrupts thinking about creation and life as the buildup toward an end product. This contrast has *ecopoethical* significance. *Ecopoethics* is a made-up term that merges ecological thinking with ethics, *poiesis* (making/creating), and poetics (language/metaphor/story). The ecological, understood as the interdependent interaction between multiple forms of life, is collective by definition. Ecopoiesis is about the many doings that cocreate, about creation as a relation of material and affective interdependency. Associating the ethical with ecopoiesis disrupts individualization and affirms ethical obligation beyond the human rational subject. Ecopoethical obligations are materially grounded in multiple living relationalities and cannot prescribe desituated norms. They are troubled by the mystery of collective material composition (for more on this conception of ethics, see Puig de la Bellacasa 2015a, 2017; for a political conception of material composition, see Papadopoulos 2018). But they oblige and have consequences. Calling upon *poetics* says that language, metaphor, and story are also consequential ecological involvements. As feminist science and technology studies show us, with inspiration from material semiotics (at the hands of thinking companions such as Donna Haraway, Susan Leigh Star, John Law, and Karen Barad, as well as emergent re-storying of naturecultures in the broad church of the environmental humanities and experimentations of literary and activist ecopoetics), it matters what stories we tell.[2] And well beyond these worlds of scholarship or even the written word, ecologies have been written into, and have inspired, affective storytelling, asking compelling questions on where and how to belong in a world (Abram 1996). The narrative in *We Are Stars*, beyond the Western scientific tropes, is that of an ontological origin tale, and like all stories in this genre, it necessarily conveys an ecopoethical account of the possibilities of agency—particularly for those that it interpellates as human, at the pinnacle of living—to participate in the productive generation of life. At stake in retelling this story, then, is how a story line that defines human belonging through elemental affinities can express the more-than-human agencies involved in the epic of life's cosmological emergence in ways that foster the actions needed to respond to Earth's current distressing conditions.

Ecopoethics is, admittedly, a mouthful: a disjointed term responding to tangled times, when ecological thinking and practices are being intensively reshaped. There is no such thing as a neutral ecology "out there" (Wright 2018). One can easily see all ecological thought as a form of hegemony, to be critically contested. Have the biological and earth sciences, by thinking of nonhuman processes as "ecosystems," not contributed to the suppression of so many other modes of inhabiting and retelling the Earth? And yet, without leaving ecolonization out of sight, in worlds dominated by technoscientific management of the environment, ecological aspirations also bear insurgent ways of rethinking ecological relationality as embedded in the doings of interdependent, more-than-human collectives that aspire to be "decolonial ecologies" (Ferdinand 2019). In the worlds I know, these ecostories are linked, for better or worse, with technoscience, and so, speaking from these impure inheritances, I feel responsible for them. These are worlds in which the imagination and material possibility of livable futures are bound with scientific stories, permeated in turn by many other ways of knowing and being in the world. Dimitris Papadopoulos (this volume) explains this condition of the ecological as "the embodied understanding of worldly connections between different beings and environments." Therefore, in the making by the interdependency of multiple agencies. This, like any intervention in the ongoing collective reshaping of "the ecological," is far from neutral. I am partial toward environmental movements—from activist protest to art interventions—crying out that the future will be ecological or will not be. I have also been shaped by specific rewriters of the "eco" (Plumwood's ecofeminism; Starhawk's "earth-based" spirituality; Myers's "affective" and "ungrid-able" ecologies) but also by those who hold the power of scientific imaginaries of more-than-human relationality into account with both love and rage (such as Haraway and Stengers). This is a situated and partial inheritance, in all its noninnocent, blind-spotted complexity. In attempting to expand ecological meanings and doings, I am no outsider observer but rather an answerable participant.

Here I try to prolong a vision emerging in recent engagements with human-soil relations that may incite ecopoethical obligations materially attuned to soils' material aliveness: specifically, to generate an obligation toward processes of *breakdown* that tempers the productive buildup of matter. We can learn from soils as media of biogeochemical relations—from how soils' biotic-abiotic multispecies communities transform matter, returning it to earth and nourishing the elemental medium for growing life again. Through these continuous, collective processes, in soils and beyond them, a myriad of organisms (bacteria, fungi, plants), imperceptibly to most human attention, not only retain chemical elements and build up compounds of matter but break them down and pass them on. Metabolizing and degrading compounds, making elements again available for reuse as nutrients and energy, soils create as they decreate. Learning from elemental breakdown in soils also compels us to delinearize the cosmological tale, to turn origins into cycles, lines into spirals, but also to complicate neat cyclic models, exposing connections, overlaps, and the multiple temporalities at play and that may alter each other. An ecopoethics embracing breakdown comes with some urgency. What does it mean to foster elemental affinities when our relations with the elementals are so messed up? If productive aliveness is a pattern of doing that forms of life share with the elementals, what other affinities can we foster to participate in the breakdown and recirculation of life?

A breakdown ecopoethics continues rather than debunks the story line of productive life in a way that acknowledges that down-to-earth life is as much the creation, the buildup of stuff, as it is its breakdown. More than a linear, future-driven accomplishment, life is an ongoing, recurrent process. The charismatic story that produces *anthropos* as an achievement of the stars would also tell us how this form of matter, and life, move on. Compounds adding to complex forms of life are meant to decompound, to recirculate, not to become stocked in perpetuating *one* particular form of life—and certainly not one that is mostly fueled by the white colonial-industrial complex, which developed a capacity to industrially manufacture compounds of matter designed to endure and resist breakdown to support its own perpetuation. It should be explicit by now how this racist-speciesist solipsism has brought ecocidal nonsense for all, including the descendants of those groups and societies that have historically retained the monopoly of livability. There is no isolated perpetuation, nor survival for a separate "humanity" in an interdependent ecosphere. Part of this is the monumental misgiving that leads to the choking of Earth's biogeochemical cycles by an excess of enduring matter that cannot break down. Concrete—along with the other synthetic, quasi-eternal compounds that have generated novel, yet artificial, "ubiquitous elements" (see Masco, this volume)—is privileged over ground.

In exploring alternative stories, I am encouraged by how relations with soils are being reconfigured, by how scientific, artistic, and activist imaginaries are contesting the productionist, exploitative, and alienating thinking about soils as a "resource" and instead promoting understandings that nurture a transformative, ecocultural resurgence of material relations and affections for the soil. In what follows, I explore the affinity of bodies with the elemental and then read this affinity through the lens of artistic visions of living bodies involved with soils in the breakdown and transformation of matter. The third section of the chapter articulates breakdown as both an ecological and ethical obligation to turn affinities with soils into solidarity with their present troubles. The chapter ends with a speculative reading of "bioremediation" technologies as noninnocent practices of ecopoethical breakdown.

Air I am
Fire I am
Water, Earth,
and Spirit I am
—Reclaiming Collective

When we gonna see that life is happening?
And that every single body bleeding on its knees is an abomination
And every natural being is making communication
And we're just sparks, tiny parts of a bigger constellation
We're minuscule molecules that make up one body
You see, the tragedy and pain of a person that you've never met
Is present in your nightmares, in your pull towards despair
—Kate Tempest

The volume *Reactivating Elements* invites ecopoethical license by embracing the elements fluidly and playing with resonances and analogies between the seemingly antithetical ontologies of elemental cosmologies and the elements of chemistry. Engaged ecopoethically, soils are elemental in all these ways. They perform in cosmological plays. In a range of languages, the word for this planet, *Earth*, is synonymous with terms for *soil* and *ground*. The sense that soil constitutes the element earth resonates too with the politics of ecological commons, such as when political ecologist Giovanna Ricoveri, in her book *Nature for Sale*, rereads Empedocles's four elements—air, water, earth (soil/land), and fire (energy)— as vital elemental commons, emphasizing them as inalienable wholes against their treatment as commodified "resources" to piece out and sell (Ricoveri 2013). However, in Western scientific language, the four elements are a relic of the time when science was closer to metaphysics and the terms referred to states of matter rather than chemical substances. Taken as a whole entity, today soil would be spoken of in science as a "natural body" or a "medium" (for plant growth, for instance). In turn, biogeochemical stories would not call soil an element: rather, the elements of chemistry and their compounds circulate through soil and contribute to making it. Soils mediate biogeochemical relations as multispecies communities that take care of different processes circulating carbon, nitrogen, and so on.

And yet, in biogeochemical stories, cosmological elemental wholes and discrete chemical meanings can become entangled in narratives that exceed scientific realms and contribute to broader transformations toward ecological cultures. This is particularly visible in commonplace story lines such as the water cycle, where water—in itself not a chemical element but a compound—is envisioned as a vital elemental medium entangled with the circulation of chemical elements such as nitrogen. The communication of ecological concerns about modifications in water chemistry becomes affected by how it upsets inalienable elementality: water as a medium for life. Similarly, plots around climate change and pollution mobilize air as elemental as it is altered by carbon dioxide, the excessive elemental compound. The *European Nitrogen Assessment,* for instance, explicitly brings together elementals and the elements of chemistry by proposing an account of "key societal threats of excess reactive nitrogen in analogy to the 'elements' of classical Greek cosmology" (Sutton et al. 2011, 92). The authors note specific chemicals that endanger the classic elements: air is menaced by nitrogen oxide, fine particulate matter, and ozone; water is threatened by nitrate and dissolved nitrogen; "greenhouse balance" (a proxy for fire) is burning with greenhouse gas and aerosols; and "soil quality" (a proxy for earth) is polluted by organic nitrogen and acidification. At the center is a "fifth" undefined element, "ecosystems and biodiversity," which depends on the balance of all the others and is threatened specifically by ammonia and organic nitrogen. In this story of the biogeochemical cycles identified by science, "analogy" is used to bring back cosmological narratives of elemental mediums to reactivate awareness and actuality about the invisible presence of noxious chemical elements around us. In this return to the elemental imaginary in scientific and policy reports and public science narratives, as well as by environmentalists calling for their protection, air, water, and soil are consistently invoked as elusive, indivisible wholes that earth creatures are made of, dwell in, and depend on for survival.

This ecocultural reactivation of elements has an ethicopolitical dimension promoted by an imaginary of elemental affinity. When I taught an undergraduate business course about the urgency of environmental issues, I realized my students became more receptive when we explored air as an elemental medium, without which life on earth as we know it would not exist. Suddenly they understood that the fate of air today (from pollution to carbon trade) was linked to the behavior of the living bodies, human and not, that depended and fed on it. By *elemental affinity,* I mean a material identification with the wholeness and inalienability of the elemental media supporting the more-than-human community that fosters a sense of belonging to earthy substances.

Finally, material affinity with the elementals (cosmological and chemical) brings origin myths, such as humans being made of mud, back into a popular imaginary of ecological cultures. These are pervasive in contemporary mangas and cartoons where both heroes and villains can embody an element or unleash its powers (for example, *Ninjago*, *My Little Pony: Friendship Is Magic*, *Avatar*, *The Last Airbender*, and *Frozen 2*, where four forces of nature are called upon as water, earth, air, and fire). There are many of these returns of the elementals today, implicit and explicit, subtle and simplistic, that deserve their own cultural unpacking in all their multiplicity and divergences.

What can we make of this potential to invoke and foster a sense of elemental affinities with earthy matter? My approach to elemental affinities as an emerging theme in the regeneration of ecological cultures in the Global North is marked by my own encounter with a path of spiritual involvement through the elementalism characteristic of contemporary Anglo-American neopagan spirituality. In particular, I am referring to the Reclaiming tradition, a Western, originally mostly white, contemporary reinvention of witchcraft, which has connections to New Age but with a strong political dimension and which is rooted in US feminist, anticapitalist, antiracist, and nonviolent earth-based activist spiritualities. Reclaiming witchcraft storytelling draws upon many eclectic sources, including modern scientific stories, old pagan lore, and acknowledged inspiration—though not unproblematic—from Indigenous knowledges and struggles. It is also strongly influenced by the political activism and spiritual writings of one of its founders, Starhawk, which are anchored on *five elements* of magic (Starhawk 1999, 2004)—the fifth being spirit. Yet to speak of spirituality here is somehow misleading, as the Judeo-Christian connotation of separating body from spirit is not present: materialism and embodiment are fundamental to this tradition, and spirit is everywhere. I have argued that material spirituality would be a better term (see Puig de la Bellacasa 2015a).

A relation of elemental affinity is expressed in the chant opening this section, which is sung at Reclaiming witch camps: "Air I am, Fire I am, Water, Earth and Spirit I am." This form of material-spiritual elementalism fosters the awareness and feeling that our bodies are composed of these elemental media/wholes and share a destiny with them: what we do with and to them will affect us, and what our bodies do, in turn, will affect the elementals. Here affinity implies sympathy, empathy, and compassion. It is another way of saying that our material constitution, as well as our actions, partakes in biogeochemical cycles. Using ritual to embrace and feel the cycles that bind us to the elements is a familiar practice in Reclaiming gatherings. The water cycle specifically was the driving story of a witch camp I attended in the summer of 2016 in Shropshire, in the UK, organized by Dragonrise, the UK branch of Reclaiming. The camp theme experimented with what Astrida Neimanis (2017) calls "being of water"—in other words, learning from the properties of water as a process of ontological reorientation of material ecological belonging. But since political activism is an important part of material-spiritual practice in the Reclaiming tradition, we also sought to translate this sympathy into solidarity with a natural body which is in trouble precisely because humans not only need it but exploit it.[3] And so we explored ways to *feel* the water's troubles through sharing stories of pollution, drought, river privatization, water injustices, and water wars, as well as more hopeful accounts of successful restoration and bioremediation of waterways. This initiation into elemental affinity is often offered as a reinterpretation of long-lost and silenced ecological relations. But for other traditions of spiritual activism, affinity and solidarity with elemental media have never been lost, even in the midst of devastation, if there is a commitment to protect the sacredness of that which is not meant to belong to people but to be guarded by them. An example is the Indigenous environmentalism of "water protectors" and other guardians of "sacred grounds" (see Gruenewald and Schmidtpeter 2015).

In common with these struggles is an urgency in the feeling—as expressed in Kate Tempest's verses above—that "every natural being is making communication," that their "tragedy and pain" are "present in your nightmares, in your pull towards despair." Ecopoethically—as a material and ethical obligation that has political but also affective/aesthetic implications—interconnectedness is key to elemental affinities. The contemporary reactivating of elemental affinity is naturecultural. It speaks of a relation that is not necessarily "chosen" or at least not in the conscious sense, as in the neoliberal sense of choice. It is driven by a "natural liking," says the dictionary. But affinity relations are not based on the usual "naturalness" associated, for instance, with ties of blood, family, or an imagined community such as a nation. Affinity is not identity either. For biochemistry, it is a biochemical "tendency" of substances to combine. And a tendency is not a given. It is changeable. Do you feel an affinity with water or with air? Do you feel flowing as a river or fiery as a fire? Are you drawn to open skies? Do you need a glass of fresh water or taking a deep breath? And, of course, affinities here inherit the sense cultivated as organizational tools in social movements (Starhawk 2002). Neither merely "chosen" nor merely given, affinities certainly need to be cultivated to flourish (e.g., by rallying around a common desire, aim, or taste). Once composed, affinities both support and oblige us. They are noninnocent, because they stir the tensions between self-interest and selflessness.[4] Who is in relations of affinity with whom and who does not, and who will be excluded as a result, remain critical questions to be asked. Elemental affinities are not, by essence, compassionate or loving; think for instance of appeals that feed nationalist exclusions by identifying people's bodies and identities to a particular soil or land (see Münster 2017; Puig de la Bellacasa forthcoming; Van Sant 2018). Specifically, elemental alliances are also the tools of lethal technocapitalist colonization. Without the mastery of chemical affinities, some wars and colonial enterprises would have been very different. Consider, for example, the relationship of nitrogen-compound soil fertilizer with war technologies (Gorman 2013), as well as what Malcolm Ferdinand (2019, 181–92) calls "the master's chemistry": the unhinged use of pesticides in the colonies, which were intrinsic to the transformation of the world into a plantation. A romanticized longing for elemental affinities would ignore that the intimate knowledge of fire has been crucial to industrialization, for the use of fossil fuels, the fractionation of oil and gas, and the formation in general of petrochemicals. Industrialization has never been dissociated from the elements.[5] So not all elemental affinities are desirable, in their specific and situated complexities. But the reactivation of the elements into a diversity of ecocultural meanings may be an opportunity to remember the importance of cultivating care and attention in relations with the powers of elements and elementals.

It seems particularly crucial, then, that elemental affinity understood as solidarity be not only about feeling with and for nonhuman forces but also about learning *from* them in different ways that disrupt reductionist technoscientific perceptions. In her essay "We Are the Flood," decolonial social theorist Claire Blencowe (2016) walks us in a landscape, inviting us to join in an attempt to listen to nonhumans. She asks, for instance, the places to "teach us" how to fight against privatization, the attack on "common life." Following cues as we wander with her (with pictures of the sites taken by Julian Brigstocke), we hit a stone wall:

Look at this wall! Look at the pipes sticking out and through it. What's it trying to do? What's it trying to tell us?

Oh I see it now! We are the flood water.

We are the water that was trapped by the culverts, forced into unnatural currents for private profit.

The flood is the struggle against the privatisation of the water.

The water is common life.

We are the flood. . . .

We are the flood. Identifying with the inanimate . . .

The differently animate?

. . . with water, helps us to think about ourselves as a collective, a collective force, rather than as individual agents.

We learn how to be a mass from matter! Mattering. (Blencowe 2016, 55–62)[6]

Canalizing water into pipes is treating water as a resource for human use, as with grand dam projects and river diversions with far-reaching effects for humans and nonhumans (from drought to catastrophic flooding and land erosion). Listening to and feeling with water, Blencowe blurs the distinction between how we treat water and people and what people and water can do. Invoking Father Jean-Bertrand Aristide's prophetic call to Haitians ("We are the flood"), she connects the message she hears from the landscape to a political struggle. Elemental affinities speak of naturecultural, ethicopolitical, more-than-human agency and *communication* (the process of making common: commoning, communing).

The question facing ecologically distressed worlds is a need to develop a sense of our inevitable "visceral entanglements" with elemental forces (Weston 2017), as well as asking what kind of more-than-human affinity clusters we need to cultivate to support each other in countering and resisting the violent endurance of lethal compounds.[7] What relations of care and obligation can be nurtured for these conditions in their situated specificities (Tironi and Calvillo 2016; Tironi and Rodríguez-Giralt 2017)? With this question I dig into the soil, as the proxy for the earth elemental. "Let there be breakdown" could be a motto for the ethopoetical elemental affinity with soil I am calling upon here—an echo of Blencowe's motto for human-water affinity, "We are the flood." These mottos are kin manifestations of specifically charged earthly powers. "Let there be breakdown"—a process that provokes control and stirs avoidance— might require embracing.

I have heard this motto resonating through visions of human-soil affinity in artistic work displaying a material-spiritual elementalism that both complements and disrupts origin stories—from stardust to life—by bringing this cosmicity down to bodily, everyday transformative processes.

Cuban-born North American artist Ana Mendieta developed a unique combination of body art with land art that is most visible in her renowned *Silueta* series. In more than a hundred iterations, she inscribed a female-shaped body on wet sands, mud, and grass by physically imprinting her own *silueta* (a Spanish word for a body figure or silhouette that is more often used for female bodies) on the ground and photographing the traces it left. By enacting ritualized encounters between her body shape and the ground as moments of transient material cotransformations, the *Siluetas* also express the inspiration Mendieta found in Cuban Santería and feminist earth-based forms of mysticism. The *Siluetas* are ephemeral, but the images capture their process of diluting, unraveling, burning, and rebecoming some other expression of life. Mendieta herself spoke of her art as "grounded on the belief in one universal energy which runs through everything; from insect to man, from man to spectre, from spectre to plant, from plant to galaxy" (Viso 2004). This sense that bodies participate in common material processes invokes the galactic elemental affinities of *We Are Stars* but brings them down to earthy changing bodies, rather than figures of transcendental attainment. This work speaks of immanent universal energy manifested in many life forms, shape-shifting through the cycles. The first photographic testimony of these encounters (*Image from Yagul*, 1973)—distinctive because Mendieta actually remains within the picture—shows her lying naked at the bottom of a Zapotec tomb in Oaxaca, in contact with bare ground and with white flowers growing from her body. Here transformation is about death, brought to the soil and transforming into life in the form of a blossoming plant: rebirth and resurrection through elemental recirculation. Mendieta's work, and the *Siluetas* in particular, have been commented on extensively (see Blocker 1999; Ortega 2004). Thinking of Mendieta's pieces as earth art, or "soil art" (Adams and Montag 2015), helps me to conjure the elemental affinity between bodies and soils that makes death inseparable from regeneration. The promise of salvation in returning to the soil requires shifting shape (see Marhöfer and Lylov 2015a, 2015b): the elemental communing of matter.

Human-soil material affinities are more explicit in a contemporary performative piece by Australian visual ecologist Aviva Reed called *Soil Biome Immersion*. This participatory "performance-lecture" experimented with a sensory approach to soil as a medium of planetary, albeit intimate and proximal, community. In an image of the performance, we see a group of people sitting around a pile of soil, all digging their hands into it.[8] In her artist web page Reed explains how she sought to "enable an immersion of oneself into the biodiverse and complex realm that is soil." She invited viewers to expand the imagination to understand the role of soil in the ecosphere as part of our "own ecological ontology." Soil, she says, "cycles nutrients temporally through the planet and hence binds all organisms to be *ancestral remnants of each other*" (emphasis added). Reed's soil biome immersion is an exercise of elemental affinity where cosmic deep time and everyday touching meet, "enabl[ing] an understanding of molecules that once resided within *the primordial soup* becoming part of one's self" (emphasis added). Reed's work explores scientific theories, retelling them to nurture an ecological imagination. Her narration mobilizes scientific plots and also resonates with the *We Are Stars* origin story. She presents us with a dazzling thought: that matter has remained somehow the "same," that Earth and its beings are still made of the bits that started it all, "ancestral remnants of each other," permanently recycled and recomposed—with the corollary that there is no waste in this system but mere recirculation of the previously combined. But again, as with Mendieta, this story suggests that the processing is not originary but ongoing and everyday. The ecopoethical consequence is that we are involved, in the present, in the continuation of these processes. Soil, here, becomes the epitomic elemental medium of this biogeochemical recirculation.

Ecopoethical affinity with soils is also evident in Claire Pentecost's motto "Our bodies, our soils." With this slogan, as part of her installation *soil-erg* at the dOCUMENTA(13) art exhibition in 2012, she invited participation in an encounter that, like Reed's, aimed at experiencing our connection with soils. Here the sense of ontological interdependency with soils becomes explicitly unavoidable—our bodies *are* soil. We can recognize a reference to the famous book title and slogan of the women's health and sexuality movement, *Our Bodies, Ourselves* (1970), that promoted body self-knowledge and self-care against the control of the medical profession. The motto promotes active engagement, rather than passive objectification. Reclaiming soils in this way restores a connection that is mediated not by expertise but by involvement. In Pentecost's work, do-it-yourself knowledge relations, with microscopes for all, intensify the proximity to soils as she invites participants to "SEE—through microscope—living beings." A broader relational engagement integrates mystical knowing (using the slogan "Composting is alchemy") and ecopolitics ("Soil is local"). As in much of Pentecost's research-based artistic interventions around soil, this work aims to help transform destructive relations with soils dominated by industrial technoscientific agriculture. Just as the notion of "one health" recognizes that the health of the environment and of humans are intimately involved, here the fate of soils—and potentially of all nonhuman beings—appear intrinsically connected to that of humans. Our bodies beyond ourselves, unbounded by human selfness, communing with the larger, more-than-human wholeness of soil matter.

Mendieta's shape-shifting, Reed's soil immersion, and Pentecost's soil embodiment are diverse but related entries into awareness of elemental affinity and solidarity with soils. Aesthetics and ethics are deeply entangled in these engagements as the poetic and sensual character of the human-soil connection is emphasized. They resonate as much with material spiritual elementalism as with notions, including scientific concepts of soil as the medium that recirculates elemental matter as nutrients for other forms of life: death as regeneration, permanent recycling, composting as alchemy (for a discussion of the "food web" scientific model of soil ecology, see Puig de la Bellacasa 2017). Soil here stands for a domestic, humbling relation with the more-than-human world. Wonder at dirt as much as stardust. Through my research, I have found numerous examples that emphasize this motif of human-soil reconnection. Soil art community events, in which people are invited to mingle and touch soils, are a powerful medium for this expression.[9] Among the different engagements at stake (Puig de la Bellacasa 2019), what is relevant here is an affinity with soils involving a community bound together by sharing the breakdown and circulation of matter. Of course, this is far from being an affectively and ethically neutral way of telling the story. This argument is situated within an attempt to contribute to the contemporary activation of ecoethical awareness of the importance of soils in technoscientific contexts through forms of material sympathy. It is one way to reengage and take responsibility at the heart of noninnocent imaginaries for stories that are re-animating soils as a ground of ethopoetical resurgence disrupting reductions to natural "resource," owned land, and nation/territory. Moreover, these stories can be seen as an intervention in the anthropocenic imaginaries, as they relinquish the identity boundaries of anthropos. They foster a notion of humans as part of a soil community who are encouraged to contribute to make soil by engaging with ecological ways of breakdown—becoming, as Donna Haraway (2016) puts it, "children of compost." This involves people as soil carers in ways that oppose the agricultural and extraction labor associated with soil as provider. In the reclaiming of an ecocommoning ethos, soils are not to be treated as a "resource." If people must engage with the cycles that both *grow* and *make* soils (Meulemans 2017), we may have to also learn to embrace our own breakdown.

THE AMBIVALENCES OF BREAKDOWN

BREAKDOWN:
- A mechanical failure; disruptive
- A failure of a relationship or system
- A sudden collapse in someone's mental health
- The chemical or physical decomposition of something
- An explanatory analysis, especially of statistics
—Oxford dictionaries online

We die so the others can be born
We age so the others can be young
The point of life is live, love
If you can, then pass it on, right?
We die so the others can be born
We age so the others can be young
The point of life is live, love
If you can, then pass it on
—Kate Tempest, "We Die"

Embracing elemental breakdown might not be easy. Step out of the cycles of ecological stories, and the notion of breaking down immediately has a different feel. Socially and culturally, breakdown is thought of as something to be avoided. A world that breaks down is not usually considered a good sign. It exposes the vulnerability of materials, of things and their worlds, of objects and infrastructures—the failure of mechanisms, relations, and systems. A car breaks down; materials slowly degrade and collapse; we suffer mental or spiritual breakdown.[10] Breakdown destabilizes, disrupts, and troubles. It also evokes simplification and reduction from the complex.[11] Thought of at this generic ethical, political, affective level, the call to embrace breakdown can be seen as a privilege of those like me, who live in relatively stable and sturdy worlds, where precariousness is perceived, credulously, as an anomaly. For many, breakdown is just the everyday norm. It can even be felt as masculinist aggression: "Let's break it all down."[12] And, in our worlds, many do not break down; they *are* broken down, so that others endure and continue to grow a "productive" life. Maybe, then, embracing breakdown is really an obligation for those who hold the (white) privilege of enduring to release capitalized life force, so that others who are deprived of it can live. Like many of the obligations of environmental justice today, embracing breakdown has to be thought of intersectionally, in an ecosocially situated way, so that it targets the breakdown of infrastructures and institutions that most exploit the material conditions of life for most humans and nonhumans on Earth.

When it comes to compounds, embracing breakdown is not even that radical. The idea that we must recirculate matter (for instance, by producing compounded materials that can be degraded) has integrated neoliberal economies of green sustainability but at the same time as efforts to counter the evils of throwaway consumption and planned obsolescence (by producing things that do not break down easily, and in this case, we need to learn to care for what we have put out there so that it lasts). A word for the future is *endurance*, referring to resilient materials, sturdiness, permanence, and robust designs and materials that "age gracefully." Breakdown can be dangerous, as some materials unleash pollutants and damaging compounds as they degrade. As proponents of the cradle-to-cradle revolution put it (McDonough and Braungart 2002), two cycles are at stake in the choreography of "sustainable living": one of endurance, with materials being reused as many times as possible, and one of dissolution and regeneration, where materials are to be degraded into biological substances. In the middle is the search for "porous" materials and urban structures sought by green engineers and architects such as Kotchakorn Voraakhom, founder of the Porous City Community (www.porouscity.org), whose childhood fascination with grass growing through cracks in her concrete playground led to her work in sustainable landscape architecture (see Sawantt n.d.; Voraakhom 2018). This tension between what can and cannot break down complicates the reduction of sustainability to efforts to make things endure into the future. However, this way of embracing the cycles can also reveal a complacent imaginary of ecological futures, where all we need are new, green designs and technologies—a proposed fix that has been widely criticized. So, while working with these contrasting feelings about breakdown, what can we learn from it as a process both to resist and embrace?

Above all, a call to embrace breakdown must be situated. I learned to be concerned about breakdown through science and technology studies work that advocates care as a fundamental more-than-human practice that keeps ecologies going and through my own involvement in calls for renewed attention to the vulnerability of materials, objects, sociotechnical infrastructures, and ecologies and to processes of care, maintenance, and repair (Denis and Pontille 2014; S. Jackson 2014; Puig de la Bellacasa 2017; Rubio forthcoming). Jackson (2014) in particular embraces breakdown as a key process of our worlds—"the world is always breaking" (223)—that can be "generative and productive," an unfolding to think from that emphasizes the vital work of ongoing repair and restoration. In prolongation of this compelling work, if I also started thinking about breakdown itself as something to allow and foster, it is because from the standpoint of the fragile worlds of the soils we have to care for, the sturdiness and endurance of manufactured stuff—and the technoscientific agility to combine elementals into matter that does not break down—pose a serious problem. Through my research on transformations in knowledge and affection for soils, including within soil science, I became aware that of all "natural bodies," soil is the one where the very notion of its formation, pedogenesis, is intimately related to a process of breakdown: a process where matter is composed over time with other, disintegrated compounds of matter. The traditional formula of soil formation summed up by the soil scientist Hans Jenny and referred to as CLORPT indicates that soil formation occurs because of **CL**imate, **O**rganic activity, **R**elief (geological), **P**arent material (that is, the rocks that specifically surround and embed soil), and **T**ime. Definitions of soil that include the dynamic liveliness of soil processes in its ontology feature breakdown as an essential combined action: "The parent material is *altered* under the influence of the climate and the early vegetation, the organic matter is *mixed* with the soil, the rock minerals are *weathered*, the organic matter is *degraded* slowly first into fresh humus, finally into carbonic acid, water, ammonia and nitrates. . . . The soil [is] thus defined . . . in accordance with the manner and length of time it had been subjected to the action of the pedogenetic factors" (Hartemink, 2016, 34; emphasis added). Scientific concepts of soil formation have not traditionally included the idea of people as "soil makers" involved in the ongoing re-creation of a soil community, but this is changing (Meulemans 2017; Puig de la Bellacasa 2017). This awareness makes one acutely sensitive to the effects of anthropogenic waste. From the perspective of soil as a living ontological entity that deserves ethical affection, even more than from the perspective of soil as a medium to produce yield, anything that humans make that cannot become soil creates bioinfrastructural issues. Making, creating, producing, maintaining, and taking care in this context are implicated less with preserving and avoiding the wearing out of things than with degrading and decomposing—and of ensuring that stuff can degrade.

It is soils that invite thinking of breakdown as an ethos. This is again present in materialist spiritualities that see soils as the places where we end—and start again. Soil is a place to think with *infranatural* forces—a word made up to invoke a below-ground mystery that does not transcend matter (Puig de la Bellacasa 2015a). In his superb classic on soil ecopoethics, *Dirt: The Ecstatic Skin of the Earth*, arborist, writer of prose and poetry, and soil eulogist William Bryant Logan (1995) describes the initiation of the degradation process of bodies. He speaks of graves as "motherly to the earth." Citing Francis Bacon's notion that "putrefaction is the work of the spirit of bodies" (54), Logan explains that *the same enzymes* that keep our metabolism regulated become "self-breaking" (56) when we die to initiate the process of returning our matter to dirt: "The soil of graves is the transformer" (57). Yet death offers a case study of people's resistance to breakdown, of those who have perfected efforts to keep the body from degrading—often using techniques similar to those used for preserving meat. Self-preservation vies against self-breaking. Logan notes, that for centuries, many of these techniques simply delayed the process, without harming soils, but other methods were more damaging, such as injecting arsenic or the use of another biocide, formaldehyde, which was introduced in the late nineteenth century.[13] Driven by the idea of protecting the living from contagion, churchyard burial grounds in societies using these methods started contaminating the broader environment: "a dead body, messy as it becomes, is not toxic in itself, formaldehyde is. It coagulates protein in the corpse and in any creature that tries to consume it" (58). Biogeochemistry was made to play against elemental breakdown. With the acceleration of chemical industrial production, technological innovation was put in the service of the cultural aversion to decay, setting the goals of endurance and self-preservation against ecopoethical sense. The living forms on Earth most resistant to breaking down—industrial peoples in white anthropocentric cultures, who lead the quest for self-preservation—excel at violently breaking "otherized" humans and nonhumans to harness or extract their biogeochemical life force (Yusoff 2018). Not only is this a horrific injustice that needs to be opposed with reparative justice, but from the perspective of biogeochemical processes and bioinfrastructures we depend on, such as soils, breaking down the privilege of endurance is a necessity. Breakdown will happen even to the most persistent and the most enduring; the ecological issues pertain to who, how, and when it will harm and/or benefit.

Breakdown is a metonymy for processes for which different scientific approaches have different words, all referring to the cycling of nutrients and energy and the entangled processes involved: biology speaks of decomposition and degradation, chemistry of cleavages and oxidations, geology of weathering. This is how compounds rebecome substances that can be passed on via the labors of a myriad of organisms and biotic and abiotic entities making biological dissociations and associations; digesting, metabolizing, and chemically reacting to each other; returning compounds back to elements, transformed as nutrients and energy available to others. The agents are multiple, and the processes are not isolated from each other. They come in all magnitudes and temporal scales, from weather to bacteria, from enzymes to airstreams, from lighting blasts to slow trickles of deep time.

In soils, breakdown mobilizes a community of beings, the "soil community"—a cooperative of organisms and plants by which elements such as nitrogen and carbon circulate and recombine. A myriad of organisms—plants, fungi, bacteria—are involved in constantly generating and breaking down compounds, retaining and passing substances around, making them available for reuse. Degradation and decomposition are processes of breakdown. Breakdown is a *relation*; entities do it to each other and sometimes for each other, giving away what they do not need anymore. It is a complex play of interdependent, albeit nonreciprocal, associations and exchanges. There is an ecopoethics at play here, something that remains difficult to grasp in its puzzling entireness, as mysterious as the vital principle was for those seeking reductionist explanations for the chemistry of life.

This approach to soil communities opens space in soil science studies for a material-spiritual dimension, one that counters the dread of breakdown as a reflection of the aversion to decay and the passing of time. The mantra "Life is death is life is death is . . ." comes to us from feminist science and technology studies scholar and anthropologist Natasha Myers. This message, which emerges from her sensory walks in the Toronto savannah forest (see Myers 2017), is an entrancing refrain of the liveliness of decay. Kate Tempest reminds us, "*The point of life is live, love / If you can, then pass it on, right?*" And as geographer Mark Jackson (2012, 207) noted while reflecting on undecomposable plastic islands, decay "as an inescapable ontological condition" has immanent ethical significance. We need, in the words of human-soil relations anthropologist Kristina Lyons (2016, 147), to think of "decomposition as life politics." It is with these kindred spirits that I embrace breakdown as an ecopoethical obligation.

But how can we foster our capacity for breakdown as an elemental affinity to all in a world stunned by the pervasive chemical compounds that will *not* break down and which are altering elemental cycles? I have never known anything other than a world permeated by the scientific, industrial, and agricultural revolutions. Most of today's dangerously unbreakable compounds will probably remain toxic as they break down again, potentially forever (see Papadopoulos, this volume). So, though an ethos of acknowledging breakdown as elemental has always existed, its revival is an ecological praxis of everyday hope but possibly without a future (Bresnihan 2018). It is this particular technoscientific timescape that I am trying to evoke.

BIOREMEDIATION AS ECOPOETHICS

Materially and poetically, ethically and aesthetically, breakdown is elemental in and for soils. It is disrupted where manufactured chemical compounds massively obstruct the capacity of soils to live, regenerate, and circulate. These compounds include hazardous chemicals, pesticides, concrete that covers breathing grounds, and other contaminants but also nitrogen fertilizers, which allegedly represent a great technoscientific achievement to which "we" all owe a large debt but which, in reality, we have *too* much of (Erisman et al. 2015; Gorman 2013).[14] From the perspective of soil care that emphasizes ecopoethical affinities, the question remains: How can an ethos of breakdown be fostered in a situated way, when the soils are not coping, not filtering, and cannot break down all the stuff we are putting in them? Thinking about breakdown in the present conditions, therefore, shows that the way I have been invoking soils as part of contemporary ethopoethical reclamations has the tone of an allegory, an idealized ecological story. Even the notion of soil being formed naturally, as described in classic soil science, seems like an archetypal figure—one with strong capacities to affect but which contrasts with the more domestic day-to-day reality that most soils on earth need help and are far from this conceptual state. Meanwhile the ecopoethical challenge is pressing: to help our naturecultural worlds to become better at breakdown; to both "dismantle" (Murphy, this volume) compounds *and* engage with the chemical regimes of scientific knowledge and everyday living that create them, rather than keeping them at a distance with critique (Papadopoulos, this volume). And for that, the soil community of breakdown needs to reintegrate people in its definition (Puig de la Bellacasa 2017). This will not be done "naturally," nor neutrally. Nurturing elemental affinities with soils ("our bodies, our soils") and creating novel ecological cultures around soils are a way of making us feel for their troubles. Becoming an ecopoethical soil champion—engaging in people-soil "conspiring" (Choy, this volume)—comes with an obligation to *assist* breakdown.

I had one of my first ethnographic confrontations with the troubles of assisted breakdown during a guided visit to an anaerobic digestate plant in Stafford, United Kingdom—a trip organized by the British Society of Soil Science as a complement to a workshop on the use of biosolids to enhance soil fertility. The plant belongs to and is managed by a family of farmers and consists of giant "digestate tanks" that break down and decompose food waste until it becomes recyclable as a liquid form of organic fertilizer. The remaining solid waste is filtered out and becomes a side product, which is turned into compost. Helping waste to break down (digesting it) aims to tackle the extensive buildup of matter by people, make something out of it that feeds the soil, and is less costly for farmers. I was told during the visit that the plant makes its money by diverting supermarket waste from the landfill and that most of the liquid fertilizer is then sold to the local farming community at a nominal price. I saw this as a good example of contemporary techniques contributing to soil making and food production while avoiding the use of industrial fertilizers.

I had attended the biosolids workshop and the corresponding tour in the hope of finding connections between scientific knowledge and alternative industrial/farming practices of fertilizing soils. On the site visit, when I looked for the area where the compost was piled, I discovered that the ground was colored with plastic confetti. The farmer told me, with patient annoyance, that much of the confetti originated in the thin plastic of yogurt containers: once shredded, she explained, "these get everywhere" and are not stopped by the digestate's filters. This compost is still used, but not for growing food. Later, back at the workshop, I overheard scientists mocking the so-called biodegradable bags in which food waste is collected in some UK towns. They had experimented with helping the bags decompose, to no avail. What stayed with me most that day was confronting the reality of contemporary soil making beyond the visions of community, activist, and domestic composting that populated my ecopoethic imaginary. That pile of "compost" saturated with minuscule bits of shredded plastic got to me. As I stood there, gauging the challenge of medium-scale, industrial breakdown practice, it became clear that most things we are building up cannot be broken down to become soil again by a return to ecological cycles. Even the semi-industrial and relatively low-tech assisted form of breakdown found in biodigestates is defeated by the endurance of compounds.

Much of the imaginary of shared matter I have relayed as the affective trigger of soil elemental affinities is based on a material-aesthetic vision of soils as kin to degrading bodies, all belonging together in a community that embraces processes of degradation and corruption of matter. But in sight of this pile of multicolored muck, this vision seems paradoxically purified. A down-to-earth, domestic, everyday connection to healing the filtering capacities of soils cannot happen today just by a reconnection to "natural cycles"; it has to involve the scientific, community, and even industrial practices of soil making.

If our bodies are our soils, and if we are to learn how to embrace elemental affinities, we must have stories that embed ecopoethical obligations in the actual material realities of soils that need assistance in breaking down matter. And then, as these stories develop, they will integrate novel transdisciplinary engagements with soils that multiply their ecological meanings. They may be about the technoscientific soil too (Meulemans 2017)—a soil whose role in the biogeochemical cycles that circulate nutrients and substances is in crisis. We need stories about soils displaced in plastic bags, from who knows where, to feed allotments and back-garden plots (Cahn et al. 2018); about soils that stink; about soils asphyxiated under concrete; about soils that make people sick; about the soil community's exploited labors (Krzywoszynska 2020). Indeed, stories increasingly are making contemporary relations with soils more complex but also more substantial and obliged (Salazar et al. 2020). These new stories put the elemental affinities of the inheritors of the agroindustrial revolutions up to the task to respond. And part of this is the need to engage with the ambivalence of the scientific enterprise in modern capitalism, which has taken much of the blame for the productionism that has choked nature while also being heralded as the promise of salvation, the path toward technologies that will solve the environmental crisis. But how could we safely assist the breakdown of compounds created with technoscientific processes and mixes without some scientific assistance? How can we locally address the global challenge of intoxicated soils without precise knowledge of the toxics underground? One task for critical friends, for those who tell scientific stories, is to continue striving to make science closer to what Dimitris Papadopoulos (2018) calls a community technoscience, a distributed form of invention power.[15]

In the search for such noninnocent, ambivalent stories, I became interested in practices that fall under the umbrella term *bioremediation*. These are incarnations of an obligation to assist breakdown that both subscribe to and trouble the imaginary of affinity with elemental breakdown and cycles. Indeed, bioremediation is consistently affirmed across the specialized literature as a range of natural processes. A classic definition reads, "Bioremediation is a *natural process*, which relies on bacteria, fungi, and plants to degrade, break down, transform, and/or essentially remove contaminants, ensuring the conservation of the ecosystem biophysical properties" (Masciandaro et al. 2013, 399; emphasis added). But bioremediation technologies are not "natural." Another definition states that "bioremediation is a branch of environmental biotechnology often used to *hasten* [the degradation of contaminants] and it *guarantees* the restoration of damaged ecosystems, using the metabolic capabilities of bacteria, fungi, yeast, algae, and microbial mats to degrade all contaminants harmful to living organisms. . . . Without the activity of microorganisms, the earth would literally be buried in wastes, and the nutrients necessary for life would be locked up in detritus" (Bonete et al. 2015, 24). Other definitions emphasize "*harnessing* the degradative potential of biological systems" (Cummings 2010, v). Bioremediation and correlative techniques (microbial and phytoremediation) are wholly naturecultural and technoscientific. However, while they involve scientific knowledge and complex techniques, they require local knowledge and strong community realization, as well as the long-term and hard work of maintenance so that the processes of cleanup are durable and not just a single event (Darwish 2013). The use of bioremediation techniques is often described as "working with nature"—a trope shared by grassroots, scientific, and industrial bioremediation narratives. *Working with* hints that bioremediation is not only a specific technique of cleaning up contaminants but a way of reintegrating practice—laboring, affective, and ethical engagements—within the natural cycles.

I cannot go into detail here about divergences and communications between community projects and scientifically oriented bioremediation approaches—for instance, the sharp contrast between the way that activist bioremediation projects affirm that "it works" and the much more hesitant and situated character of bioremediation technologies (which do not always translate well from lab to field), as well as issues with a functionalist and economizing reduction of organisms to their "role" in performing ecosystems services. But that bioremediation is an imaginary inseparable from the technoscientific realm is incontestable. This does not mean ignoring the contrasts between industrial clean-up-and-leave approaches—which focus on removing the contaminant—and the more ecosystemic engagements of local communities with the long-term needs of healing and restoring. What can we learn about embracing breakdown by staying with these tensions? What kind of imaginary of science-based ecological obligation do these elicit? Two notions appear to me in the way ecological obligation is asserted in these different enactments.

In a speculative reading, bioremediation is not just a scientific term but an ecopoethical doing that can be embedded in practices of care and obligation, as well as material-spiritual, more-than-human collective healing. Bioremediation as ecopoethics features a set of hopeful technologies based on a sense of ecological commitment to *heal life with life* across very different spaces, processes, affections, future imaginaries, and fields: scientific, activist/community technoscience, artistic and aesthetic, and technoindustrial. And yet, engaging speculatively with bioremediation is an opportunity to complicate and transform the stories of "restoring" human participation to biogeochemical cycles. This is precisely because bioremediation comes from *inside* predominant plots of renaturalization of elemental circulation in the so-called natural cycles. Bioremediation is as ambivalent as any technologically mediated close affinities with "nature." This ambivalence cannot be made less ambiguous if an affirmative breakdown ecopoethics is to be possible. The fragile promise of ecological salvation in these technologies, as well as their perilous craft, reveals a mitigated, more domestic reality. The breakdown of compounds is a recirculation, including toxic recirculation. These micro-experiments in earth chemistry are not a panacea: they might release elements back into widely disrupted cycles, and degrading compounds can contribute to further elemental excess and atmospheric intoxication as the released substances are put into recirculation (see Papadopoulos and Puig de la Bellacasa forthcoming).

Finally, bioremediation is an ecopoethic trope for anthropocenic times, because it faces contamination and ecological disruption from the perspective of the *aftermath* of ecocatastrophe—what Murphy (2017, 497) calls the alterlife, or "life already altered, which is also life open to alteration." Bioremediation activates the imaginary of a world that needs to be regenerated and repaired, suggesting that we can work with nature for its own salvation.[16] It is an imaginary of ecological obligation within the acceptance that the ecocatastrophe is not to come, it is here; it has been happening for a long time for most people and will continue unfolding (Beuret 2015). Assisting breakdown is an obligation stemming from the acceptance that one particular catastrophe has already happened: the excess of manufactured compounds that is choking ecosystems. Here the task of elemental affinity politics is to start learning to break all this stuff down again by turning origins into returns and foster ecological belonging through cultivated ecopoethics. That means assisting breakdown by any means possible and, in the process, creating something else that already embraces the capacity of breakdown. In any case, the elemental affinities we need right now cannot be promoted by stories that reduce people's participation to ecopoesis to the buildup of matter. They belong to alternative imaginaries of life, not as an achievement but as ongoing and plural collective creation-decreation.

AFTER STORIES: NAUSICAÄ AND THE SCIENCE
OF NATURE THAT HEALS ITSELF

And what do good stories do? . . . They kill you a little.
They turn you into something you weren't.
—RICHARD POWERS, *The Overstory*

This chapter started with a story and ends with another one, told by masterful storyteller Hayao Miyazaki ([1984] 2012) in his stunning two-volume manga *Nausicaä of the Valley of the Wind*.[17]

In the far future, humankind has destroyed the Earth in a war known as the Seven Days of Fire. This devastating, technoscientifically driven chemical war left behind the Sea of Corruption, the name given to a forest of mold that now engulfs great parts of Earth. This forest is actually a giant fungus that emanates thick clouds of spores which humans cannot breathe—a miasma forcing everyone to wear masks in most places. The respiration of a forest is poisoned through roots and soil. There is also a proliferation of giant insects, the Ohmu, which humans both fear and respect and whose enormous lightweight shells they use for construction materials. The plot is complex, full of mythologies, religions, and tribal and dominion powers. But what interests me is the role of scientific knowing and how the manga plays with the idea of a nature that heals itself. I read this tale as a parable of the tentativeness and ambivalence of science-based ecological remediation, as part of the struggle to reconfigure more-than-human relations in the midst of ecological devastation.

The sciences and technologies that are said to have brought destruction are controlled, impossible to use again, or buried in crypts. Small pockets of human collectives survive, but most are at war with each other under the control of tribal warlords. One such pocket, trying to remain at peace against all odds, is the people of the Valley of Wind. Their young princess, Nausicaä, is a salvation figure. Loved by all, she is the chosen one who was prophesied to save Earth and humanity. We learn that Nausicaä is secretly an amateur, self-taught scientist, and rather than fearing and trying to destroy the toxic jungle, she tries to understand it. She has constructed a garden lab in the deepest cellars of the castle that she keeps secret in order not to scare her people. Here she takes spores she collected from the Sea of Corruption and grows them in clean water and clean soil. She even grows a hisokusa, one of the most dangerous plants. She discovers that the spores do not create miasma when raised with clean water and clean soil, and she concludes that the Sea of Corruption is actually slowly removing the poisons from the earth.

At the same time, Nausicaä is a warrior who must confront a new war coming up between tribes. One of the tribes has resuscitated biological warfare, and with the help of remaining scientists, they have recovered old, malevolent biotechnologies that were used during the Seven Days of Fire.

Science, as the story unfolds, has two radically opposed sides: manipulation for power, dominance, and senseless destruction and Nausicaä's experimental science, dedicated to understanding the forest and driven by care, empathy, and love. Nausicaä's influence comes from her skills as a warrior, which have been adapted to her environment, but also from the fact that she loves everyone and everything, even the giant insects and the fungal growth that scare everyone else. She even feels the suffering of the brutal warriors leading the destruction, who are destroying themselves in the process. It is through this love of all things that Nausicaä has managed to investigate and understand the toxic forest and other life forms that she lives with. Her journey seeds peace across a destructive landscape of war but is also one of research, thoughtful observation, and sensing what this forest is and how it might have come to be, what its relation might be to the past destruction, and how it might hold a future for Earth's beings. Crucially, she questions what possible role humans can play in the remediation of a world that has become uninhabitable for them, without eradicating the strange living beings that now populate it and thrive in it. In her quest, she discovers a hidden scientific people below the forest who have also been monitoring the toxic jungle and have understood that the evolution of this fungal and vegetal overgrowth represents a slow process of healing, cleaning, and purification. She learns that these people have accepted that the time of healing will last for hundreds of years, but rather than attempting to harness the process, they see themselves as witnessing and assisting it without hindering this remediation, by which life is breaking down the toxics into a harmless powder. In sum, they are involved; they see themselves as part of the forest's process, rather than as the main agents of its remediation.

But just when the story seems to meet the soothing and hopeful ending of letting nature work, it takes an unexpected turn. Nausicaä learns that the forest of corruption and the large insects are actually forms evolving from a past biotechnological manipulation, programmed by scientists who, after the Seven Days of Fire, realized what humanity had done and put their science to work to create a naturecultural technology that would ultimately clean Earth again. She discovers that she herself is a scientifically modified human, created to survive in the atmospheric miasma until the purification is complete. More disturbing still is that humans will eventually need to be altered again, to be able to breathe newly purified air. Once she realizes this, she decides to interrupt the program in order to escape the alternative between purity and corruption. At the conclusion, we do not know if and how Nausicaä and her people will survive. The story ends setting humans back in an impure present, as modified forms of life accepting the obligations of living with the ancestor technosciences that have produced them and their environment but without accepting the destiny they had been built to demarcate. There is no hopeful ending—no hints of lasting peace in the aftermath of ecocidal war—but not one of despair either.

Read as an allegory of bioremediation in the aftermath of catastrophe, this tale disrupts the imaginary that situates hope in the longing for clean worlds or self-healing ones, whether brought by nature or by science. Assisting breakdown will require embracing scientific practices, but these might have to go by many other names too. The tough ecological obligation that Nausicaä incarnates is to engage with inherited worlds, formed by past and complex mediations, to love them and care for them. Yet at the same time, another hesitant promise for technoscience's inheritors engaged in assisting the breakdown of compounds lies in a multitude of diverse and situated minor remediating doings, many of which are not waiting for those who call themselves humans to know or care but require to be let to be.

ACKNOWLEDGMENTS

The paper that became this chapter was first presented in September 2016 at the annual meeting of the Society for Social Studies of Science in Barcelona. Since then I have presented versions of this work in a number of seminars and conferences, and I would like to thank everyone who generously provided comments, questions, suggestions, and critiques on each of those occasions, as well as my coeditors of this volume and anonymous reviewers for their careful readings.

NOTES

1. Pop science mobilizes the sublime simplicity of this idea. As I finalize this chapter, a tweet from @latestinspace reposted by a Facebook friend refers to Carl Sagan with the same thought: "The Nitrogen in our DNA, the calcium in our teeth, the iron in our blood, the carbon found in our apple pies, were made inside collapsing stars." As Sagan famously stated in the *Cosmos* television series, "We are made of star stuff."

2. An example of activist ecopoetics is the Dark Mountain Project. My interest in poetics as a spiritual approach was first sparked by Susan Leigh Star's and Geoffrey Bowker's evocation of a "poetics of infrastructure" (see Puig de la Bellacasa 2015). This led me first to work by Stuart Cooke and then to classics such as Gander and Kinsella (2012). While my knowledge of this field is clumsy and superficial at best, I thank my University of Warwick colleague Jonathan Skinner, editor of the journal *Ecopoetics*, for generous suggestions to continue this exploration and engage with forms beyond narrative, including poetry on soil and decomposition.

3. Our work at the camp was connected to Earth Activist Training, a collective of courses that combine permaculture, earth-based spirituality, organizing, and activism: http://earthactivisttraining.org.

4. I thank Natasha Myers for drawing my attention to this tension, about the ways in which "allies" might be motivated by self-interest while a shift toward a language of "accomplice" could point to other forms of solidarity and affinities: http://www.indigenousaction.org/accomplices-not-allies-abolishing-the-ally -industrial-complex.

5. I thank an anonymous reviewer for prompting me to emphasize this aspect.

6. Blencowe (2016, 67) connects the message she hears from the landscape to the decolonial political fight of the Haitian left-wing movement known as Lavalas, a Haitian Creole word for *avalanche* or *flood*. She quotes party leader Jean-Bertrand Aristide: "Alone we are weak, together we are strong, altogether we are the flood. Let the flood descend, the flood of poor peasants and poor soldiers, the flood of the poor jobless multitudes. . . . And then God will descend and put down the mighty and send them away, and He will raise up the lowly and place them on high."

7. Treating alliances between humans and nonhumans in their specificity rather than as the incarnation of many ontologies is at the heart of Papadopoulos 2018.

8. Aviva Reed, Soil Biome Immersion, http://www.avivareed.com/#/soil-biome -immersion, accessed April 19, 2021.

9. Many of these events appeal to sensual affection for soils. They include creative workshops conceived as artistic/performative/community events that invite people to play, touch, and feel soils, such as Naomi Wright's *Soil Kitchen*; "soil-tasting" sessions, in which participants smell different local soils placed in wine glasses and then taste food grown in that soil, such as Laura Parker's *Taste of Place*; "Dirt Don't Hurt" meditation sessions, where participants sit on soil-filled pillows or sleep with a test tube filled with soils from different locations tucked under their pillow and record their ensuing dreams; Amanda White and Alana Bartol's interventions as part of their Deep Earth Treatment Centre project; and events that embrace the sexual appeal of soils, such as the "Wedding to the Dirt" ecosexual performances that involve marriage rituals as well as rolling naked in the mud (in Elizabeth Stephens and Annie Sprinkle's SexEcology work). All of these interventions incite material engagements to cultivate our commonalities with soil substance.

10. While I cannot explore this aspect of the discussion here, there are ways in which a mental health breakdown can also be a breakthrough—something I heard uttered in the compelling 2017 film *Crazywise* by Phil Borges and Kevin Tomlinson. This is not just a metaphor: when things are too much or people become enmeshed in life situations that cancel all possibilities—or exhaust them—a breakdown can be our only way out. I also feel this speaks well of the current state of ecologies choked by materials that are blocking the living flows.

11. Breakdown also invokes a simplification—as in the analytical breakdown of a data set or an aggregate—that requires a rupture of links and relations. We break down larger wholes and complexities to make them more approachable, to make them easier to assimilate, to bring incomprehensible scales to thinkable ones—similar to the way guts break down chunks of food to be processed. Breakdown is about things that become smaller particles, from the compounded into the elemental, about dissolving, about rebecoming dust.

12. Thanks to Nerea Calvillo for pointing out this connotation.

13. Thanks to Natasha Myers for pointing out that formaldehyde has many other uses and an interesting history, which Nick Shapiro has extensively written about.

14. The justification for putting nitrogen-based fertilizers into the soil is that they are needed to feed a growing population. Agricultural productionism has placed soils on steroids. And yet more than 60 percent of the fertilizer that goes into the soil is not taken up by crops and instead runs off too fast for the organisms that recirculate these compounds (the denitrifiers) to do their work. Thus it seeps into waters, together with sewage that has been improperly disposed of, and creates the well-known phenomenon of eutrophication, or dead zones in waters. Nitrogen compounds are now seen as a necessary "good" thing that has become "too much" as the cycles are overwhelmed. Science-based policy recommendations mostly tackle the source, in particular reducing meat consumption. I hope to be proven wrong and see it in my lifetime, but it seems an impossible task to cut use of nitrogen-based fertilizers while the World Bank (2013) is encouraging market expansion into "underfertilized" zones. Astronomical amounts of capital have been mobilized and ride on the untouchable moral argument that we need to feed ten billion people. We are stuck with nitrogen compounds; their widespread use seems unstoppable.

15. Besides the examples given by Papadopoulos (2018), inspiration can be taken from work in which the boundaries of activism, scholarship, and science blur. What Wylie et al. (2014) refer to as "civic technoscience" takes many forms, in projects such as Public Lab (https://publiclab.org/), the Environmental Data and Governance Initiative (https://envirodatagov.org/), Write2Know (http://write2know.ca/), and the Endocrine Disruptors Action group (https://endocrine disruptorsaction.org/).

16. There is also the suggestion that bioremediation can help us save the world, as evoked by the title of a celebrated book by the mycologist extraordinaire Paul Stamets (2004), who teaches about mycoremediation (bioremediation that mobilizes mushrooms and fungi to clean contaminants) in grassroots permaculture trainings.

17. I am immensely grateful to Anna Krzywoszynska for telling me that I must absolutely read this.

REFERENCES

Abram, D. 1996. *The Spell of the Sensuous: Perception and Language in a More-Than-Human World*. New York: Vintage.

Adams, C., and D. Montag. 2015. *Soil Culture: Bringing the Arts Down to Earth*. Devon, UK: Centre for Contemporary Art and the Natural World.

Beuret, N. 2015. "Organizing against the End of the World: The Praxis of Ecological Catastrophe." PhD diss., University of Leicester.

Blencowe, C. 2016. "We Are the Flood." In *Listening with Non-Human Others*, edited by J. Brigstocke and T. Noorani, 53–72. Lewes, UK: ARN Press.

Blocker, J. 1999. *Where Is Ana Mendieta? Identity, Performativity, and Exile*. Durham, NC: Duke University Press.

Bonete, M. J., V. Bautista, J. Esclapez, M. Camacho, J. Torregrosa-Crespo, R. M. Martínez-Espinosa. 2015. "New Uses of Haloarchaeal Species in Bioremediation Processes: Advances in Bioremediation of Wastewater and Polluted Soil." In *Advances in Bioremediation of Wastewater and Polluted Soil*, edited by S. Naofumi. London: InTech Open Access.

Borges, P., K. Tomlinson. 2017. *Crazywise*. Seattle: Phil Borges Productions.

Bresnihan, P. 2018. "Hope without a Future in Octavia Butler's *Parable of the Sower*." In *Problems of Hope*, edited by L. Dawney, C. Blencowe, and P. Bresnihan, 39. Lewes, UK: Authority Research Network.

Cahn, L., C. Deligne, N. Pons-Rotbardt, N. Prignot, A. Zimmer, B. Zitouni. 2018. *Terres des Villes: Enquêtes Potagères de Bruxelles*. Paris: Editions de l'éclat.

Cummings, S. P. 2010. *Bioremediation: Methods and Protocols*. Totowa, NJ: Humana Press.

Darwish, L. 2013. *Earth Repair: A Grassroots Guide to Healing Toxic and Damaged Landscapes*. Gabriola, Canada: New Society.

Denis, J., and D. Pontille. 2014. "Material Ordering and the Care of Things." *Science, Technology and Human Values* 40 (3): 338–67.

Dominguez, Rubio F. Forthcoming. "On the Discrepancy between Objects and Things: An Ecological Approach." *Journal of Material Culture*.

Erisman, J. W., J. Galloway, N. B. Dise, A. Bleeker, B. Grizzetti, A. M. Leach, W. de Vries. 2015. *Nitrogen: Too Much of a Vital Resource*. Zeist: World Wildlife Fund Netherlands.

Ferdinand, M. 2019. *Pour une écologie decoloniale: Penser l'écologie depuis le monde caribéen*. Paris: Editions du Seuil.

Gander, F., and J. Kinsella. 2012. *Redstart: An Ecological Poetics*. Iowa City: University of Iowa Press.

Gorman, H. S. 2013. *A Story of N*. New Brunswick, NJ: Rutgers University Press.

Gruenewald, T., and L. Schmidtpeter, dirs. 2015. *Sacred Ground* (documentary film). sacredgroundfilm.com.

Haraway, D. 2016. *Staying with the Trouble: Making Kin in the Chthulucene*. Durham, NC: Duke University Press.

Hartemink, A. 2016. "The Definition of Soil Since the Early 1800s." *Advances in Agronomy* 137:73–126.

Jackson, M. 2012. "Plastic Islands and Processual Grounds: Ethics, Ontology, and the Matter of Decay." *Cultural Geographies* 20:205–24.

Jackson, S. J. 2014. "Rethinking Repair." In *Media Technologies: Essays on Communication, Materiality and Society*, edited by T. Gillespie, P. Boczkowski, and K. Foot, 221–39. Cambridge, MA: MIT Press.

Krzywoszynska, A. 2020. "Nonhuman Labor and the Making of Resources: Making Soils a Resource through Microbial Labor." *Environmental Humanities* 12:227–49.

Logan, W. B. 1995. *Dirt: The Ecstatic Skin of the Earth*. New York: Norton.

Lyons, K. 2016. "Decomposition as Life Politics: Soils, Selva, and Small Farmers under the Gun of the U.S.–Colombia War on Drugs." *Cultural Anthropology* 31:56–81.

Marhöfer, E., and M. Lylov. 2015a. *Shape Shifting*. Berlin: Archive Books.

Marhöfer, E., and M. Lylov, dirs. 2015b. *Shape Shifting* (film).

Masciandaro, G., C. Macci, E. Peruzzi, B. Ceccanti, S. Doni. 2013. "Organic Matter–Microorganism–Plant in Soil Bioremediation: A Synergic Approach." *Reviews in Environmental Science and Bio/Technology* 12:399–419.

McDonough, W., and M. Braungart. 2002. *Cradle to Cradle. Remaking the Way We Make Things*. New York: North Point Press.

Meulemans, G. 2017. "The Lure of Pedogenesis: An Anthropological Foray into the Making of Urban Soils in Contemporary France." PhD diss., University of Aberdeen.

Miyazaki, H., dir. (1984) 2012. *Nausicaä of the Valley of the Wind*, deluxe edition. San Francisco: VIZ Media.

Münster, D. 2017. "Zero Budget Natural Farming and Bovine Entanglements in South India." *RCC Perspectives: Transformations in Environment and Society* 1:25–32.

Murphy, M. 2017. "Alterlife and Decolonial Chemical Relations." *Cultural Anthropology* 32:494–503.

Myers, N. 2017. "Ungrid-able Ecologies: Decolonizing the Ecological Sensorium in a 10,000 year-old NaturalCultural Happening." *Catalyst* 3 (2): 1–24.

Neimanis, A. 2017. *Bodies of Water: Posthuman Feminist Phenomenology*. London: Bloomsbury Academic.

Ortega, M. 2004. "Exiled Space, In-Between Space: Existential Spatiality in Ana Mendieta's *Siluetas* Series." *Philosophy and Geography* 7(1): 25–41.

Papadopoulos, D. 2018. *Experimental Practice: Technoscience, Alterontologies and More-Than-Social Movements*. Durham, NC: Duke University Press.

Papadopoulos, D., and M. Puig de la Bellacasa. Forthcoming. "Eco-Commoning in the Aftermath: Sundews, Mangroves and Swamp Insurgencies." In *Swamps and the New Imagination: On the Future of Cohabitation in Art, Architecture, and Philosophy*, edited by N. Urbona and G. Urbona. Berlin: Sternberg.

Puig de la Bellacasa, M. 2015. "Ecological Thinking, Material Spirituality, and the

Poetics of Infrastructure." In *Boundary Objects and Beyond: Working with Leigh Star*, edited by G. Bowker, S. Timmermans, A. E. Clarke, et al., 13–46. Cambridge, MA: MIT Press

Puig de la Bellacasa, M. 2017. *Matters of Care: Speculative Ethics in More than Human Worlds*. Minneapolis: University of Minnesota Press.

Puig de la Bellacasa, M. 2019. "Reanimating Soils: Transforming Human-Soil Affections through Science, Culture and Community." *Sociological Review* 67:391–407.

Puig de la Bellacasa, M. Forthcoming. *When the Name for World Is Soil*.

Ricoveri, G. 2013. *Nature for Sale: Commons Versus Commodities*. London: Pluto.

Salazar, J. F., C. Granjou, M. Kearnes, A. Krzywoszynska, M. Tironi. 2020. *Thinking with Soils: Material Politics and Social Theory*. London: Bloomsbury Academic.

Sawantt, S. n.d. "Kotchakorn Voraakhom: Fighting Climate Change." Rethinking the Future. Accessed August 4, 2020. https://www.re-thinkingthefuture .com/know-your-architects/a980-kotchakorn-voraakhom-fighting-climate -change.

Shapiro, N. 2015. "Attuning to the Chemosphere: Domestic Formaldehyde, Bodily Reasoning, and the Chemical Sublime." *Cultural Anthropology* 30:368–93.

Stamets, P. 2004. *Mycelium Running: How Mushrooms Can Help Save the World*. Berkeley, CA: Ten Speed Press.

Starhawk. 1999. *The Spiral Dance: A Rebirth of the Ancient Religion of the Great Goddess*. San Francisco: HarperSanFrancisco.

Starhawk. 2002. *Webs of Power: Notes from the Global Uprising*. Gabriola, Canada: New Society.

Starhawk. 2004. *The Earth Path: Grounding Your Spirit in the Rhythms of Nature*. San Francisco: Harper.

Sutton, M. A., G. Billen, Albert Bleeker, Jan Willem Erisman, Peringe Grennfelt, Hans van Grinsven, Bruna Grizzetti, Clare M. Howard, and Adrian Leip. 2011. *European Nitrogen Assessment: Technical Summary*, xxxv–li. Cambridge: Cambridge University Press.

Tironi, M., and N. Calvillo. 2016. "Water and Air: Territories, Tactics and the Elemental Textility of Urban Cosmopolitics." In *Urban Cosmopolitics*, edited by Anders Blok and Ignacio Farías, 207–24. London: Routledge.

Tironi, M., and I. Rodríguez-Giralt. 2017. "Healing, Knowing, Enduring: Care and Politics in Damaged Worlds." *Sociological Review Monographs* 65:89–109.

Tsing, A. L. 2016. *The Mushroom at the End of the World: On the Possibility of Life in Capitalist Ruins*. Princeton, NJ: Princeton University Press.

Van Sant, L. C. 2018. "'The Long-Time Requirements of the Nation': The US Cooperative Soil Survey and the Political Ecologies of Improvement." *Antipode* (November).

Viso, O. M. 2004. *Ana Mendieta: Earth. Body Sculpture and Performance, 1972–1985*. Vienna: Hatje Cantz.

Voraakhom, K. 2018. "How to Transform Sinking Cities into Landscapes That

Fight Floods." Filmed November. TED video, 12:22. https://www.ted.com
/talks/kotchakorn_voraakhom_how_to_transform_sinking_cities_into
_landscapes_that_fight_floods?language%20february%2011%202019#t-9373.

Weston, K. 2017. *Animate Planet: Making Visceral Sense of Living in a High-Tech Ecologically Damaged World.* Durham, NC: Duke University Press.

World Bank. 2013. *Growing Africa: Unlocking the Potential of Agribusiness.* Washington, DC: World Bank.

Wright, W. J. 2018. "As Above, So Below: Anti-Black Violence as Environmental Racism." *Antipode* (September).

Wylie, S. A., K. Jalbert, S. Dosemagen, M. Ratto. 2014. "Institutions for Civic Technoscience: How Critical Making Is Transforming Environmental Research." *Information Society* 30:116–26.

Yusoff, K. 2018. *A Billion Black Anthropocenes or None.* Minneapolis: University of Minnesota Press.

10

EXTERNALITY, BREATHERS, CONSPIRACY: FORMS FOR ATMOSPHERIC RECKONING

Tim Choy

ELEMENTARY ENVIRONMENTAL ECONOMICS

In economics, an "externality" is a cost that is not accounted for within a particular order of pricing calculation. Represented as a graph, it looks something like figure 10.1. The graph is a classic. I first encountered it more than twenty-five years ago, but it continues to be taught and used today. At the time I learned it, I was an undergraduate student studying "earth systems," a new major being developed within Stanford's College of Earth and Geological Science. Thinking back, I see earth systems' establishment as an early moment in the institutionalization of a mode of interdisciplinary planetary systems thinking that aspired to an integrated appreciation of systems spanning life and physical sciences, one designed to foster a nonarbitrary interdisciplinarity for aspiring environmental scientists. Students would develop core competencies and exposures in key fields—namely, chemistry, biology, physics, geology, and economics—while also electing a track, or "sphere" of earth systems,

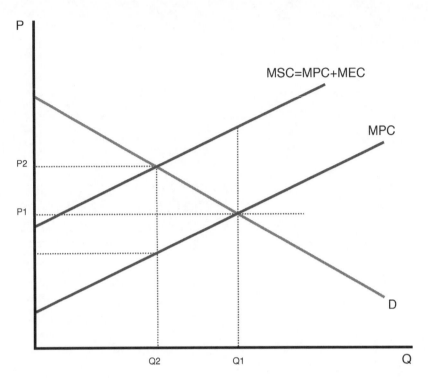

FIGURE 10.1: Environmental externality graph

in which to focus. As a student of "biosphere," I studied evolutionary ecology and conservation biology but also interned in a US Geological Survey laboratory, processing paleolithic materials to construct data on historical climate change. "Geosphere" students might model geochemical cycles that transported carbon, nitrogen, and other elements in not only atmospheric and lithic but also organic forms. The "anthrosphere," meanwhile, referred almost exclusively to economics and policy.

The disciplinary contours of what might be legible as the anthrosphere within earth systems became clear to me after a failed petition to substitute an alternative to the core environmental economics requirement. I had proposed to pursue study with a literature professor on historical narratives of nature, natural scarcity, and the racialized figure of the human as a biological organism, but the proposal agitated my earth systems adviser, a prominent evolutionary ecologist. "Some people want to say there is no scarcity!" the biologist sputtered. "That's nonsense!" It was 1992, not yet the heyday of the so-called science wars, but my request touched a nerve made sensitive by early forms of environmental denialism. In response, my adviser thickened the line drawn between the realms of science and narrative, creating in our advisory conversation an unbreachable wall between environmental science and questions concerning culture and cultural assumptions. While there might be human sciences within earth systems, they would not include study of the culture, discourse, or narratives of the earth systems sciences themselves.

So I enrolled in my required introductory course in environmental economics with trepidation. Like many in my biophilic cohort inspired by deep ecology within mainstream student environmental activist circles, I was skeptical of the assumption that it was necessary to assign human value to environments and environmental changes to mobilize environmental concern. I dreaded learning the practice of assigning monetary value to life-forms and ecologies.

But early in the term, I learned of this elementary graph, and it rocked my world—though perhaps not as my instructor intended. The graph may be familiar to some readers, but a detail of the way it was taught stood out to me, and I want to highlight it here, for it merits thinking about slowly.

The graph represents an externality. Here, *externality* means the costs that are not paid if environmental and environmental health impacts are not taken into account. The line aiming down from left to right is known as the demand curve (D). The demand curve's downward slope as it moves from left to right abstracts a conditional relation where a small quantity of goods will be demanded (bought) if the price is high, while larger quantities will be purchased if the price is low. Now notice the darker lines heading upward from left to right; these represent "marginal costs" for the producer. Costs (of production) increase as more goods are produced. Finding the point where the light gray demand curve and the dark gray cost curve intersect is supposed to yield the ideal price and number of goods to make.

The lower of these two dark lines is the conventional cost curve; if one works with that line, the ideal price, the point of intersection between cost and demand, is "P1." This intersection of the demand and cost curve is known as the equilibrium point, a place of ideal pricing where the producer will not leave any profit on the table or lose money from overproducing.

"But what if in fact there are more costs?" the economist asked as he drew and explained the graph. He described environmental costs, social

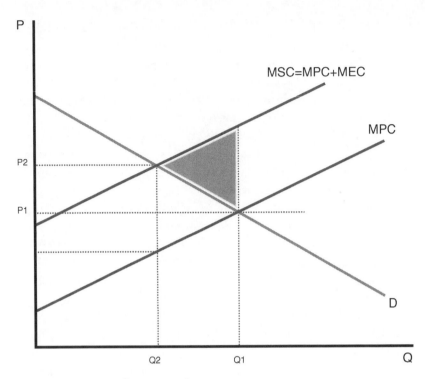

FIGURE 10.2: Externality as triangle

costs—costs that are not paid directly by the producer. "The real cost curve, then, would be here," he said as he drew the second, higher, diagonal line. (The fact that there are two lines here indicates a difference between the [marginal] personal cost [or apparent cost] and the [marginal] social cost [real cost] of producing a good.)

This means the most economically efficient price, the "equilibrium point," is actually at P2, where the light gray demand curve meets the second, higher, dark gray cost curve. The lower price, P1, makes sense only when the social costs are not accounted for.

The social costs that are not paid when price does not reflect the real costs can be visualized rather strikingly. In the graph, it takes the form of a triangle, as the area between the two cost curves and between the quantities of goods produced in the two pricing scenarios. You will have to imagine, with a flourish, the blackboard work. First the economist draws the standard demand and cost curves. Then he adds the social cost. And now, he turns the chalk and shades in the triangle (figure 10.2).

"Here is the unpaid cost: the externality," I remember him saying. "These are real costs, but they are *externalized*, meaning they are not paid."

"But someone does pay," he continued. "Who pays?"

He waited a beat, then answered himself. "*Breathers* pay."

BREATHERS PAY: BREATHERS AS EXTERNALIZED ELEMENT

Breathers pay. In this context of an environmental economics lecture, where the point was to introduce the concept of an externality and a mathematical means to visualize and calculate it, *breathers* served as a variable to refer to a whomever—an unspecified collective of human or nonhuman subjects who would bear the brunt of an impact in an abstract scenario. This would then motivate the curriculum for different, better economic calculations. Accounting for health care costs if production processes cause environmental health problems, or learning ways to estimate the lost value of ecosystem services when a landscape becomes degraded for the production of a good—these would shift pricing upward so the prices of goods would reflect their true costs. *Breathers* was a metaphor.

But I could hear the phrase only literally. *Breathers pay.* What could this mean, and what did it do to say it? For years, since that day in the lecture hall, I have been grappling with the repulsions and pulls of that utterance, and I would suggest that even as the concept of an externality has become conventional and a matter of course, if one keeps lively the notion of breathers paying, then the figure of breathers does more work than simply being a variable for abstract social cost. The concept's viscerality is an opening to more than law and economics. Who are breathers? What does it mean to say that they pay? What did it mean, and what *else* could it mean, to bring breathers and breathing into account in the calculations of costs and benefits?

This chapter is a speculative reflection on forms and experiments in atmospheric reckoning. *Reckoning* is a dense word. My use of it learns from Diane Nelson's (2009) study of sense-making practices emergent in the wake of civil war and violence in Guatemala. Nelson ethnographically gathers forms of record keeping, joke telling, and scandal, characterizing them as modes of reckoning with an immensity of violence that is hard to grasp. In doing so, she leans into the word's simultaneous evocations of both accounting and accountability—senses of both a keeping count and a settling of affairs. Reckoning signals here at once a recording of extent and a pointing to a direction forward. In this chapter, I am interested in forms of "atmospheric reckoning," by which I mean efforts to account for and to come to grips with, through both technical and nontechnical means, the diffusions and concentrations of airs and surrounds, including the ways conditions of atmospheric experience distribute uneven life. This mouthful is intended as a modulation, not a displacement, of environmental justice. Its fixation on air and breathing is meant to contribute techniques for specifying modes of effect and relation that are proper to airy relations and dependencies within multiscalar, multiphased, unequal lifeworlds in formation.

To return to the graph at the beginning of this chapter, I consider the concept/figuration of externality itself a form of atmospheric reckoning insofar as it bodies forth a calculus and geometry for taking into account that which is not accounted for before. This reckoning's form makes a certain demand. I am interested in asking what that form of reckoning does and what other forms for atmospheric reckoning might be activated from within its logics and terms. My concern with forms of reckoning activated by the charge *Breathers pay* thus takes me somewhat to the side of the question of what the elemental is and whether air (or breath) is an element or ensemble. My interests are rather in what different modes of reckoning with breath, breathing, and breathers might do for a speculative project of collectivizing response to massive, patterned forms of environmental violence.[1]

CONCEPTS, METHODS, THEORY

This chapter works narratively and imagistically. I narrate a conceptual iteration, one signaled through transforming concept images. Specifically, this chapter offers a punctuated movement in thinking and working with atmospheric subjection. Its signal terms—*externality, breathers,* and *conspiracy*—mark shifts in concept and concern: from the accounting of things unaccounted for, to the speculative work of aggregating a concrete vocative, to the ethicopolitical question of what collectivities might be catalyzed or mediated by sense of atmospheric complicity.[2]

Figure 10.3 offers a guide: I begin from a graph that visualizes a staple of environmental economic thought, the externality. I extract a shape and concept from it in order to ask how else it might be populated. This will lead me to a concept figure, the Museum of Breathers, that I have

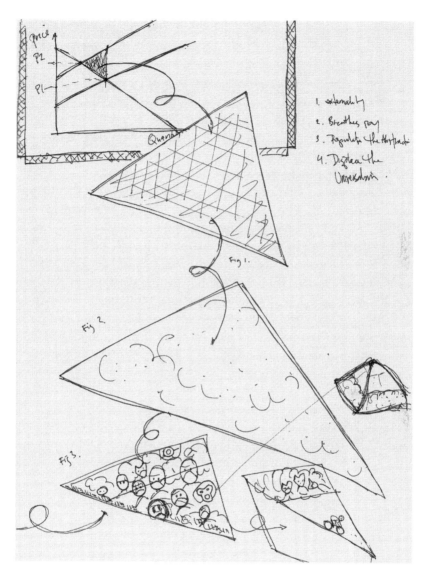

FIGURE 10.3: A map diagramming a movement from externality as abstraction to aggregation of breathers and conspiring collectivity.

found useful for a speculative gathering of breathing subjects and different breathing conditions meant to pose an unequally shared cosubjection. This will prompt a movement from questions of cosubjection to the implications and demands of collectivity—in other words, the implications and demands made today in unequally shared conditions and demands of breathing together, or conspiracy.

I adopt a narrative form and include images to make explicit the processual back end of how I have come to ask questions of atmospheric problems in particular ways. Through studies of how people work with smoke, spores, soot, and smells, I have been working through concepts such as suspension, distribution, concentration, and reduction, as well as breathers and conspiracy—experimenting to see if they bring me to a different idiom for thinking about relations. To some audiences, these concepts feel abstract; but while they are abstractions, they are not ideal in the sense of deriving from thought, nor are they distant from the "real world." Instead, these abstractions are highly specific, concrete concepts, tools for opening a field for knowing and thinking atmospherically by taking the cues from qualities of atmospheric things, while also accounting for their manifestation as such (that is, as atmospheric) through particular apparatuses of attunement of particular atmospheric sciences (see Choy 2018).

Generally, I work through attention to qualities or conceptualizations of relation emergent in different people's ways of engaging with air and breath. I then ask what other forms might be within reach to open another mode of thinking from within the constraints and invitations of those qualities and concepts. I often draw things encountered during research when I reach a limit with writing (i.e., when I find myself writing in the same way). I am an amateur drawer, not an artist, and the considerable effort it takes to render an image leads me to linger with details I might not have noticed before, details that can yield differently grounded concepts. In turn, I write when I reach a limit with image making. This is not unlike the way the aesthetic theorist and filmmaker Eric Cazdyn (n.d.) has described his process of filmmaking and writing—though Cazdyn's films show more skill than my drawings!

One could think of this as working formally. I would just note this is formal work that follows the form of materials and relations given by atmospheric phenomena, rather than forms given in advance, presumed, or chosen arbitrarily. Another way to put this is that I work through speculative entailments of forms that I encounter in the course of atmospheric study. *Entailment* is an old linguistic/logical term that refers to a condition where a proposition can be construed to follow logically from what comes before. When I say "proposition," I mean it in the sense that Isabelle Stengers (2005) has advanced, in the sense of a proposal, an invitation to a different contour for thought (for more on the difference between writing to characterize or propose and writing to prove, see Stengers 2015, 33–34). A speculative entailment, then, is a proposal for another line of thought that could follow what is given—if different details in the given were sensible, sensed, and activated.

FROM EXTERNALITY TO BREATHERS

What, then, does an externality entail? The concept can point in a number of directions in economic theory. In its early conception, in the work of Cambridge economist Arthur Cecil Pigou (1877–1959), it was tied to the idea of offset taxes. Pigou built on his teacher Alfred Marshall's initial concept of external costs to argue in *The Economics of Welfare* (1920) that where certain costs imposed on others are not accounted for by the person who creates these costs, they should be offset by a tax, so that the real costs might modulate the degree of that economic activity. Later, Ronald Coase (1960) argued that economic solutions more efficient than taxes might be found if property rights were distributed to involved parties, including those who were not accounted for previously, and transaction costs reduced so that the parties might bargain with each other for the rights to continue or stop that activity. Economic historian Deirdre McCloskey (1998, 367) characterizes one thrust of Coase's argument: "The 'theorem' is supposed to be that it doesn't matter where you place the liability for, say, smoke pollution, because in a world of zero transaction costs the right to pollute will end up in the hands that value it the most. If breathers value it most they will buy it. If steelmakers value it most they will keep it." In this mode of thought, a solution without a "Pigovian tax" and its implicit ascription of corporate blame might be found if—and only if—all of the affected parties can be identified, all effects can be quantified, property rights and entitlements can be found or made and distributed correctly, fair tables for negotiation can be established, and costs of exchanging land or breathing rights for money can be eliminated. It is this *aspirational* sense of externality as potentially internalizable—in other words, an idealized, frictionless social plane where the market could internalize an externality on its own—that undergirds the hope currently placed in carbon offsets, carbon markets, and the trade of rights to pollute.

More could be said about how Coase's argument for the superiority of a market solution for internalizing an externality came to inform what became known as the "Chicago School" of law and economics, but that is beyond the scope of this chapter. Here, my aim is to point out how the externality, in its basic definition, posits that there are agencies and effects that emerge as an outcome of, yet outside the scope of, transactions between the usually recognized actors in an economic scenario. It is here, in the triangular space opened by the externality, that "breathers" can enter the scene as affected parties and potential transactors.

Furthermore, that the concept of the externality so fundamentally structures environmental economics today is perhaps unsurprising, facilitated by Pigou's extended discussions of the unaccounted-for social costs and benefits caused by such diffuse things as air pollution, smoke abatement, and lighting (see Pigou [1920] 2017, part II, chap. 9), as well as by Coase's (1960, 1) prefatory remarks in his response to Pigou that "the standard example [of those actions of business firms which have harmful effects on others] is that of a factory the smoke from which has harmful effects on those occupying neighbouring properties." The economic externality was thus from the outset a reckoning with what, following media theorists such as John Durham Peters (2015) and Nicole Starosielski (2015), we might term the elemental media and infrastructures of a particular mode of economic life.[3] Viewed in this light, the externality concept is both a moment and a form of atmospheric reckoning. By this, I mean that the concept emerged as a response to a need to reckon with airborne effects and that the externality offers itself as a structure and logic of value for that reckoning.

Still, it bears mention that the externality and the externalized triangle are less explanations than they are characterizations. As concept figures, they work as prompts rather than explanations. They do not define; they ask instead to be populated and enfleshed; they bring and beg for more. What if, through the invitation of the viscerality of the breather as idea, we gather outside the lecture hall to shift attention from the externality to the breather, to ask who breathers are and how they pay? If the externality asks for a quantification of the externalized costs of these effects, what of those subjects caught or breathing in the medium of the effects? This moves us from a project of developing better equations for costs and pricing to a practice of elaborating what is implied when breathers pay the unpaid debts in an atmospheric system.

To dwell in and to imagine what it means to say "breathers pay" means to grapple with the thought that someone's breathing—or respiratory impairment—is part of the cost. Those who inhale the exhausts of current arrangements bear a disproportionate load, not just in a particular moment of exposure but in the ongoing accumulation of body burdens that are at once chemical and historical (Agard-Jones 2013, 2014).[4] "Breathers pay" conjures a currency problem, for the phrase is chillingly polyvalent. In one reading it nominates an abstract class of affected parties who will transact money. In another, it connotes breathing itself, or its impingement, as the medium of payment. Both senses present themselves at once. Furthermore, the phrase begs a broader question: Who is *not* a breather, among the aerobic living in an atmospheric condition?

The apparent universality of the figure of the breather is dangerous. *Breather* is a vacuous term—typically only used in environmental economics to gesture to an abstract notion of social cost—and it threatens to generalize a breathing experience/position when everything that environmental justice activists and scholars have done points to a stratified distribution of concrete breathing experiences.[5] This is something to avoid.

Breathers might instead be more fruitfully employed like Marx's *workers*: a vocative meant to figure a role in a particular kind of relation, one occupied in common by multiple forms of human and nonhuman life. While its gathering function works through a gesture of indifference to particularity, the very broadness of breathers as a concept draws attention to the divergent conditions of forms of breath—gasped, wheezed, sniffed, raspy, unnoticed, worried over, meditated upon—and the impossibility of generalization.

The commons of breathing, in other words, is not one of equivalence but of partial connection, uneven distribution, and potential resonance.[6] As the calculation of externalities and neoliberal solutions for internalizing them through the market have become commonplace modes for imagining and conducting environmental management, I have felt compelled to stay with the breathers and air conditions that saturated the very conception of the environmental externality, asking what other kinds of atmospheric reckoning they might yield. I am asking whether and how such a concept and figure might be recuperated and populated to figure worlds of conjoined but unequal fate. Imagine it as filling the externality's triangle with breathers and their conditions, rather than a shade of color that denotes a quantity of unpaid costs. "People are struggling to breathe, and more so in some places than others," writes Kim Fortun, in a compelling call for science and technology studies work to develop methods of accounting for corporate or structural violence. Invoking the example of Perry, Alabama, where coal ash from a sludge spill in Kingston, Tennessee, was transported for disposal and storage and became toxic dust, Fortun continues, "There is always a Perry, Alabama, even in philosophical anthropology, marginalities that the rest depends on" (2014, 326). Breathing is always of an atmospheric uncommons; commoning it takes and does work that is both analytic and political.[7]

THE MUSEUM OF BREATHERS

For some time, my method of thinking such disparateness together and casting atmospheric relations into the foreground has been a device that I call the Museum of Breathers (figure 10.4).

It began accidentally several years ago. I had run into an impasse while trying to write about a moment when an atmospheric scientist told me of an experimental solution he and a colleague had developed for proving the link between juvenile asthma and air pollution. "You'll like this: we've finally got it," he said. The answer was so simple, he could not believe they had not thought of it before. "We use *baby* rats," he said, just

FIGURE 10.4: The Museum of Breathers in booklet form

like that—emphasis on *baby*, none on *rats*—which told me something even as it told me that I should have known it all along. Of course, there was rat model research on respiratory illness. The researchers' innovation came not from the use of a rat model, but from the use of neonates. My colleague then described for me a collaborative experimental system. The first researcher, an engineer, built an internal combustion engine. The second raised baby rats near the engine's exhaust pipe to see whether and how particular particles affected the rats' lung development.

I was struck by many things. I was struck particularly by the poignant predicament where a politics of environmental justice, itself constitutively sensitive to the unequal distribution of breathing conditions, depends upon an apparatus that cultivates unhealthy exposures into its evidentiary systems; it needs vulnerable breathers of its own. I set out to write about how to hold together and in tension the relations of debt and gift emergent in such different, and yet also similar and wholly materially and politically entangled, forms of atmospheric vulnerability. Yet I was stymied. One difficulty was speed. The two worlds I wished to hold together—that of the exposure experiment and that of the health politics of asthma and environmental justice—were both so complex. If I described one in what felt like sufficient detail, I was pages in before I could introduce the second. Yet, if I wrote them together right away, it seemed to do both of them a disservice of reduction.

At one point, to give myself an anchor while writing, I sketched a two-panel comic (figure 10.5). The sketch pairs a vulnerable child with an inhaler with two anthropomorphized murine characters in diapers. Channeling Haraway's meditations on OncoMouse, the nonhuman test subjects are labeled Asthma Mouse (a careless species error on my part!). One of them wears an air-filter mask. The neonatal mice and the human child occupy separate panels, but smoke drifts between and connects them. Their bodies and stances are similar. In the child's panel, the smoke's source is not clear. In the other, the smoke spews from a nondescript machine called Internal CombustX 2000. A caption reads, "Our lives are comingled in breath and smoke and soot. And strangely, rats poops. The rats' labored breath/wheeze—in tandem with the carburetor—make a promise of justice and an unpayable debt."

I hoped the sketch would serve as a kind of emulation target, but my paragraphs never achieved the economy and immediacy of the image. So, I turned to drawing more, populating an imaginary Museum of Breathers.

FIGURE 10.5: Drawing breath together

The Museum of Breathers is a speculative space (figure 10.6). In it, museumgoers encounter and reflect upon the entangled conditions of diverse beings offered as atmospheric kin (figures 10.7, 10.8, and 10.9). Drawn together through the curatorial conceit of the museum are asthmatic subjects in polluted sacrifice zones, neonatal rats whose bodies substantiate the risks of air pollution for the studies crucial to the politics of evidence in environmental justice activism, mushrooms tested for their capacity to metabolize and remediate radioactive fallout, equipment for atmospheric sensing (figures 10.10 and 10.11), and practitioners of breath work.

The Museum of Breathers was meant to elicit a question of what becomes possible to think about through a recognition of drawing breath

I IMAGINE IT AS A TASTEFUL, UNDERSTATED BUILDING.

A LITTLE OUT OF THE WAY MAYBE.

INSIDE, IT'S DARK. LIKE THE NATURAL HISTORY MUSEUM.

AND YOU WALK PAST THESE DIORAMAS. THEY'RE LIT, SO THEY ILLUMINATE OUR FACES AS WE WALK BY, AS WE LOOK, AS WE PAUSE.

FIGURE 10.6–10.7: The Museum of Breathers, pp. 2–3

WHAT WOULD IT MEAN, OR DO, TO FACE OUR ATMOSPHERIC KIN, WITH WHOM WE SHARE AN UNEVEN, UNEQUAL MEDIUM?

LIVING, RESPIRING, TRAFFICKING IN SUSPENSIONS. LABORED BREATHING, BREATHS AS LABOR. THEY ARE OUR CONDITIONS OF VALUE.

FIGURE 10.8–9: The Museum of Breathers, pp. 4–5

FIGURE 10.10–11: The Museum of Breathers, pp. 6–7

together and the partial connections of atmospheric practices and substance. An obviously curatorial space, the museum figure makes explicit that its organizing concept is not meant to define the world but to offer a characterization that brings a mode or sensibility through which other thoughts might arise.[8]

I drew the Museum of Breathers through serially encountered, captioned static scenes because while I was not trying to make a narrative, I sought more than the frissons of juxtaposition. Its style of drawing is figural, but far from photo-realistic. There was, I hoped, the possibility of the sensation of a concrete, aggregating, and variegating abstraction. I provided benches for sitting with particular breathers and situations.

This was a way of slowing down, while also speeding thought across the array.

LIMITS TO THE MUSEUM

There are limits to the museum. Here, it is an exceptional space of pause where a viewer, a visitor, might become transformed through an object or curated set of objects. This ascribes, or at least implies, significant authority for the curator. This is somewhat intentional; it is proper, I think,

COULD WE HOLD THESE
BREATHS TOGETHER?

COULD WE DRAW
BREATH TOGETHER?

COULD THE BREATHERS
OF THE WORLD CONSPIRE?

FIGURE 10.12: The Museum
of Breathers, p. 8

to foreground the generative work of selection and assembling and to make explicit that the concept of breathers (or a collective thereof) is an authorial assertion.

More challenging are the limits that come with building a thought space through the collection of scenes from disparate locations. By this, I mean more than the way this structure repeats a historically power-laden form of anthropological knowledge making and encounter, though I suspect that colonial legacy pertains to the conceptual limit that concerns me. This conceptual limit is that while the Museum of Breathers speculatively actualizes an ethicopolitics of facing, where presumably that facing potentially calls forth a responsibility (or, following Haraway, a capacity to respond, response-ability), it does not in itself offer much for thinking about the *stance* of such a response. Even if atmospheric others are presented as kin with each other and oneself, linked through the diffuse and enduring substance of that which is breathed, in what mode are these relations to be pursued?

Other questions follow: What happens next? What would an avowal of such relations look like? What would it take to be good atmospheric kin? To broach such issues—to move from the situation of encountering the face of an other to a project of breathing together—requires a departure from the museum. Questions of capacity and potential follow as you leave (figure 10.12): "Could we hold these breaths together? Could we draw breath together? Could the breathers of the world conspire?"

If a speculative museum seemed necessary at first as a way to gather and reflect on what it might mean concretely to say that breathers pay the externalized costs of the present, in the last decade or so, images of breaths and air conditions have saturated the news and other media (figure 10.13). They include police deployments of pepper spray for atmospheric crowd control in Davis, California, in 2011 and elsewhere (for a compelling history of the transformation of tear gas from a wartime weapon to policing technology, see Feigenbaum 2017); an explosion at the Chevron refinery in Richmond, California, in 2012, caused by pipe corrosion; Eric Garner's choking death in 2014 at the hands of New York City police officers, George Floyd's death at the knee of Minneapolis police, the Black Lives Matter (BLM) movement, and solidarities forged between BLM and the Dakota Access Pipeline protests; the Umbrella Movement and ongoing democracy struggles in Hong Kong; a rise in black lung disease in Appalachia and China; the My Right to Breathe movement in Delhi, India; a worldwide rise in asthma rates and deaths related to air pollution; and increasingly frequent wildfires and smoke in Northern California and Australia—not to mention global worries in 2020 about COVID-19's degrees of aerosolization and avenues of airborne transmission.[9]

I invoke the surge of such moments not to argue that they point to the politics of breathing. The point is not to have an elemental theory of everything. The point is rather to ask *what demands these different mo-*

FIGURE 10.13: Outside the museum

ments make of each other, when brought together, and of any effort to think the politics of breath. For me, they prompt a turn in emphasis from the moment of a gathering together of subjects by virtue of their breathing to the hard question of what collectivity and politics would be if routed through a sense of what it means and takes to breathe alongside and with respiring others in an unequally shared medium. *Breathers* as a signal *not of universality, but of complicity.*

Conspiracy. The word's Latin roots *con* ("with") and *spirare* ("breathe") could not be more literal, and using this word to figure the joint between politics and breathing brings tones of necessary intimacy and risk to the figuration of political collectivity. "Conspire means to 'breathe together,'" shares political performance artist and theorist L. M. Bogad (n.d.). "Conspirators huddle together, quietly and cautiously sharing the intimacies of breath and ideas. Conspirators confide, sharing faith and trust in a dicey environment."

Inspired by Bogad, I hear *conspiracy* as a provocation and response to the "dicey" conditions under which "breathers pay." Conspiracy compels me not only because it bears the meaning of *breathing together* in its etymology but also because the word carries a semantic and affective freight of subversive togetherness, a coming together within and against compromising conditions. These "dicey environments" are ambient, material, and political; and in the huddle is where conspiracy asks for something more than recognition, allyship, or collaboration.

Together, these semantic and figural loads bring a sense of risk to the concept image of breathers reckoning with the material conditions in which, against which, and for which they might conspire. They ask what are the "costs" paid by certain breathers or externalized by others, within a particular composition of conditions. This means a departure from the reflective space of the Museum of Breathers and a turn in emphasis from the notion of a breathing subject (and the work of eliciting a recognition of cosubjection) to a question of what it would take to think, feel, and do an ethical commitment to conspiracy—even while carrying the lesson from the museum that the sociality of breathing (and therefore breathing with) is an always more-than-human and not entirely elective formation.

Conspiracy implies a commitment to breathing together from and in an unequally shared milieu, an unevenly constituted planetary medium for respiration where concentrations of well- and unwell-being accumulate differentially, sometimes quickly, sometimes slowly. If we think conspiracy literally, what political forms might transpire from an assembly caught in and metabolically dependent upon an atmospheric uncommons? How can we think the form and formation of this kind of collectivity?

One response to these questions is offered by anthropologist and science and technologies studies scholar Kristen Simmons (2017) in her incisive theorization of "settler atmospherics." Simmons posits the surrounds of settler atmospherics as the always ambient conditions of Indigenous and anti-Black strangulation in the US settler state, one conspicuous condensation of which was the federally sanctioned deployment of tear gas and pepper spray against water protectors and their supporters at Standing Rock in 2016, yet which also permeate everyday life for Indigenous and Black people in less dramatic forms. "Breathing in a settler state is taxing," Simmons deadpans. Within and against the surrounds of settler atmospherics, Simmons takes conspiracy to figure a potential decolonial form of atmospheric reckoning: "We need to conspire to strategize logics of agitation, which displace and unsettle. Doing so calls us not to ignore difference, but to create alter-relations with one another. . . . What would it take for individuals to reconceptualize the embeddedness in which all already are with and have the potential to be for—to stage the grounds for a collective reimagining, a conspiration, an atmospheric otherwise?" Here, conspiracy does not shy from difference but rather aspires to activate an unsettling of the present toward different relations and a different world, "an atmospheric otherwise." Such conspiration might be glimpsed already, Simmons argues, in the coalescence of new solidarities, such as the Black Lives Matter national chapter's 2016 statement of solidarity with water protectors at Standing Rock: "New ecologies are forming for liberation in a settler atmosphere."

In conversation with Simmons, Sefanit Habtom and Megan Scribe (2020) have recently called for a "co-breathing" of Black, Indigenous, and Black-Indigenous peoples in North America, situating their call not only in the long history of settler colonialism in the United States and Canada but also in the wake of police killings of Black people and Indigenous peoples. These eruptive violences punctuate an ongoing pandemic present where COVID-19's impacts in North America are experienced disproportionately by Black and Indigenous people, due to lack of resources and systemic precarity of employment and housing. "Settler atmospherics might include the gas and chemicals that were used to break up Standing Rock and inflicted upon Minnesota demonstrators,

FIGURE 10.14: Dissolve

the asphyxiation of Eric Garner and George Floyd, and the respiratory impacts of COVID-19 that are felt most significantly in our communities" (2). Cobreathing, argue Habtom and Scribe, offers one way to think Black-Indigenous relations and futures in settler atmospherics with more intimacy than the language of coalition affords, attending both to the "shared experience of surviving within white settler society, while at the same time, taking seriously the antiblack and genocidal imperatives that mark us differently" (1).

These formations shape conspiracy as a gulp of the possibilities and impossibilities of being with, within, and against capitalist and colonial worlds constituted through naturalized and ambient stratifying externalities. Not an atmospheric crowd in general, but a commoning-in-action in the cloudy uncommons.[10] To conspire is to avow embodied complicities and intimacies, both those activated in breathing together and those circulating as the surround. It is to pose historical and future-oriented questions of the conditions that sustain or deplete you: What conditions the differential distribution of the difficulties or impossibilities of breath for particular forms of life here in this atmosphere? What will it take to face and alter these conditions of distribution where I inhale the cough or choke of another?

Working within the "ecological obligations" (see Papadopolous, this volume) of breathing together, conspiring collectivity brings a dense repertoire of affective entanglements. While empathy and sympathy reach across a prior distance of understanding or feeling, conspiracy owns a differentiating consubstantiation that is already commingled and in process. In this thickened attention to relations between body and breath, it is kin to what María Puig de la Bellacasa (this volume) calls elemental affinity. At the same time, as a response to the logic of the externality, conspiracy emphasizes less the fostering of an atmospheric "sense of belonging to earthy substances" (Puig de la Bellacasa) and more the ethical and political relation one makes with the elemental breathing relation of another. Think conspiracy as a breathing affinity group obliged by the condition of being already cosuspended in an unequally viscous medium of violence where breathers pay. The triangle of social cost dissolves (figure 10.14). The externality was never external, just as lungs, whose interiors circulate the atmospheric surround, are never truly internal. In its place, breathers conspire.

NOTES

1. I am compelled by this idea of reckoning as I consider the graph here as visualization and eventually create images of my own, because it offers a detour around the common arguments—made as much by scientists and artists as by critical commentators—about the function of science and/or art in making the invisible visible, making the insensible sensible, or representing the unrepresentable. Such characterizations seem unnecessarily tied to a logic of detection or metaphysics of presence. Similarly, one could argue that the recent ubiquity (and fundability) of ecologically oriented art and art practices indicate an acknowledgment of the limit of science-as-usual's techniques of representation, communication, and visualization—a limit of their capacity to elicit effect and affect—and that explicitly aesthetic practices, by which I mean practices self-consciously seeking to effect new modes of feeling and sensation, offer a necessary toolkit for reformatting the terrains of sensibility and, hence, politics. I think this latter

argument is right, but I am less interested in making the general argument and more interested in the conduct and efficacy of specific efforts.

2. As I broach the question of atmospheric collectivity, I am indebted to Cori Hayden for many discussions of crowd theories, old and new, and the poetics through which they dissolve atomic individuals into masses, swarms, waves, and more (see Hayden, this volume).

3. On earthly elements as forms of media, see Peters (2015), as well as Jussi Parikka's (2015) argument for earth itself (and its minerals) as media. Also see Nicole Starosielski's (2015) exemplary study of the undersea cabling network—and various geopolitically situated struggles that conditioned its extensions and installations—that enables the internet, as well as her recent work on heat and temperature control as infrastructural (2016).

4. Anthropologist Agard-Jones (2013, 2014, in preparation) revoices the idea of body burden from toxicology and activism, reformatting it to convey toxic accumulations that are at once chemical and more than chemical, entangling as they do the chemistry of insecticides and fungicides with unfoldings and endurances of colonial violence. Durham (1969) was one of the first to use the term in toxicology, to convey the "storage" and "relatively greater persistence" of chlorinated hydrocarbon insecticides such as DDT in US humans, relative to other kinds of chemicals, due to their fat solubility—or more importantly, their insolubility in water. A high body burden marks persistence of exposure and/or a slow chemical half-life.

5. I have many guides in this, including Kim Fortun (2001, 2014) on the uneven present of late industrialism's atmospheres; Michelle Murphy (2011) on distributed reproduction and chemical infrastructures along the Saint Clair River; Jerry Zee (2017) on the political conditioning of saltating sands in the Tengger Desert; Alison Kenner (2018, 2019) on differential emplaced embodiments of asthma and breath in Philadelphia; Nicholas Shapiro (2015) on the gray-market dispersal of formaldehyde off-gassing trailers provided by the US Federal Emergency Management Agency (FEMA) in the wake of Hurricane Katrina; Joe Masco (2015) on the fallout of engineered worlds; the Black Lives Matter movement, catalyzed in 2014 by Eric Garner's plea that he could not breathe before his death in a choke hold at the hands of New York City police officers and now resurgent as I revise this chapter in 2020 in the wake of George Floyd's choking by Minneapolis police; and Hong Kong's burgeoning democracy movement, weathering tear gas and rubber bullet storms from the Hong Kong police.

6. The term *partial connection* is borrowed from Marilyn Strathern (2004), who develops a mode of analysis where cross-context comparisons do not presume a master scale, such that the connective insights made through them are necessarily understood as both partial and constituted through the drawing together of disparate situations and scales.

7. On the commons as uncommons, see Marisol de la Cadena (2015). For de la Cadena, such a commons marks constitutive divergence such that any commons is "constantly emerging from the uncommons as grounds for political negotia-

tion of what the interest in common—and thus the commons—would be" (7). My take on commoning is indebted to Dimitris Papadopoulos's (2012) theorization of worlding justice as a political, practical, and collective work of alterontological attention to the potential but not guaranteed commoning of material worlds.

8. Atmospheres seem amenable to experiment lately, especially art/science experiments such as Beatriz da Costa's *Pigeon Blog* and Tomas Saraceno's *Aerocene* project. Many of these seem to be experiments in making air and its qualities become present or in calling forth an explication of the surrounds of ecopoliticization. Derek McCormack (2018) offers a particularly helpful take on aesthetic atmospheric experiment in his account of balloons as devices for experimenting in inducing and probing atmospheric variation, where balloons invite a sense of an envelope, a surround, an interiority and exteriority, questions of buoyancy, and more. As experimental devices, they enable sensitivity to certain qualities that matter emergently in the context of particular atmospheric interventions.

9. The extent and complexity of the planetary air pollution crisis is breathtaking. Kim Fortun's (2001, 2014) work to develop concepts, methods, and collaborative platforms for addressing its conditions, reach, and potential for dislodging is crucial. Also see the Asthma Files research project (https://theasthmafiles.org).

10. Deep thanks to Cori Hayden for conversations and collaborations that have helped me to think crowd and cloud formations together—through their overlaps, disjunctions, and tropes-in-transit between the two (see Hayden, this volume).

REFERENCES

Agard-Jones, Vanessa. 2013. "Bodies in the System." *Small Axe: A Caribbean Journal of Criticism* 17, no. 3 (42): 182–92.

Agard-Jones, Vanessa. 2014. "Spray." Somatosphere, May 27. http://somatosphere.net/2014/spray.html/.

Agard-Jones, Vanessa. In preparation. "Body Burdens: Toxic Endurance and Decolonial Desire in the French Atlantic."

Bogad, Larry. n.d. "Conspire." Accessed September 1, 2020. https://www.lmbogad.com/links.

Cazdyn, Eric. n.d. "Blindspot Variations." Accessed May 5, 2018. http://www.ericcazdyn.net/home/videosincense/the-blindspot-variations-2/.

Choy, Timothy. 2018. "Tending to Suspension: Abstraction and Apparatuses of Atmospheric Attunement in Matsutake Worlds." *Social Analysis* 62 (4): 54–77.

Coase, R. H. 1960. "The Problem of Social Cost." *Journal of Law and Economics* 3:1–44.

De la Cadena, Marisol. 2015. "Uncommoning Nature." *E-flux Journal* 65 (May–August).

Durham, William F. 1969. "Body Burden of Pesticides in Man." *Annals of the New York Academy of Sciences*. 160 (1): 183–95.

Feigenbaum, Anna. 2017. *Tear Gas: From the Battlefields of World War I to the Streets of Today*. London: Verso.

Fortun, Kim. 2001. *Advocacy after Bhopal: Environmentalism, Disaster, New Global Orders*. Chicago: University of Chicago Press.

Fortun, Kim. 2014. "From Latour to Late Industrialism." *HAU: Journal of Ethnographic Theory* 4 (1): 309–29.

Habtom, Sefanit, and Megan Scribe. 2020. "To Breathe Together: Co-conspirators for Decolonial Futures." *Yellowhead Institute Policy Brief*, June 2. https://yellowheadinstitute.org/2020/06/02/to-breathe-together/.

Kenner, Alison. 2018. *Breathtaking: Asthma Care in a Time of Climate Change*. Minneapolis: University of Minnesota Press.

Kenner, Alison. 2019. "Emplaced Care and Atmospheric Politics in Unbreathable Worlds." *Environment and Planning C: Politics and Space*, 1–16.

Masco, Joseph. 2015. "The Age of Fallout." *History of the Present* 5 (2): 137–68.

McCloskey, Deirdre. 1998. "Other Things Equal: The So-Called Coase Theorem." *Eastern Economic Journal* 24 (3): 367–71. www.jstor.org/stable/40325879.

McCormack, Derek P. 2018. *Atmospheric Things: On the Allure of Elemental Envelopment*. Durham, NC: Duke University Press.

Murphy, Michelle. 2011. "Distributed Reproduction." In *Corpus*, ed. M. J. Casper and P. Currah, 21–38. New York: Palgrave Macmillan.

Nelson, Diane M. 2009. *Reckoning: The Ends of War in Guatemala*. Durham, NC: Duke University Press.

Papadopoulos, Dimitris. 2012. "Worlding Justice/Commoning Matter." *Occasion: Interdisciplinary Studies in the Humanities* 3 (1).

Parikka, Jussi. 2015. *A Geology of Media*. Minneapolis: University of Minnesota Press.

Peters, John Durham. 2015. *The Marvelous Clouds: Toward a Philosophy of Elemental Media*. Chicago: University of Chicago Press.

Pigou, Arthur. (1920) 2017. *The Economics of Welfare*. London: Routledge.

Shapiro, Nicholas. 2015. "Attuning to the Chemosphere: Domestic Formaldehyde, Bodily Reasoning, and the Chemical Sublime." *Cultural Anthropology* 30 (3): 368–93.

Simmons, Kristen. 2017. "Settler Atmospherics." *Fieldsights*, November 20. https://culanth.org/fieldsights/1221-settler-atmospherics.

Starosielski, Nicole. 2015. *The Undersea Network*. Durham, NC: Duke University Press.

Starosielski, Nicole. 2016. "Thermocultures of Geological Media." *Cultural Politics* 12 (3): 293–309.

Stengers, Isabelle. 2005. "The Cosmopolitical Proposal." In *Making Things Public: Atmospheres of Democracy*, ed. Bruno Latour and Peter Weibel, 994–1003. Cambridge, MA: MIT Press.

Stengers, Isabelle. 2015. *In Catastrophic Times: Resisting the Coming Barbarism*. London: Open Humanities.

Strathern, Marilyn. 2004. *Partial Connections*. Updated ed. Savage, MD: Rowman and Littlefield.

Zee, Jerry C. 2017. "Holding Patterns: Sand and Political Time at China's Desert Shores." *Cultural Anthropology* 32 (2): 215–41.

11

REIMAGINING CHEMICALS, WITH AND AGAINST TECHNOSCIENCE

Michelle Murphy

On rainy days in downtown Toronto, or Tkaronto in Mohawk, the water drizzles down the glass and steel surfaces of the multinational banks spiking into the sky above the financial district. The water runs off the hard surfaces of the buildings, pouring into the storm sewer system and then washing out into the vastness of Lake Ontario, one of the five Great Lakes. The office towers are covered in a biofilm, a thin greasy layer that airborne particulate pollution and persistent organic pollutants stick to (Andre Simpson et al. 2006). Persistent organic pollutants, such as polychlorinated biphenyls, or PCBs for short, travel the globe in plumes of pollution, colliding into office towers that act like massive collection devices. The patter of rain on urban glass rinses the collection of particulate matter and industrial chemicals into a concentrated wash that flows down into the urban sewers. returning water and chemicals to the lake that is itself a legacy dumping ground of PCBs from an era of twentieth-century settler colonial industrial exuberance.

Lake Ontario is the last Great Lake in the eastward flow of water. From west to east, the lakes become more industrialized and engineered. As water flows out of Lake Huron into the international shipping channel of the Saint Clair River, it passes the petrochemical corridor of Canada's "Chemical Valley" and then moves through the (de)industrialized zone of Detroit to join Lake Erie, which is rimmed by the Rust Belt stretching from Ohio to New York, was declared "dead" in the 1960s, and is regularly smothered by algae blooms. The water continues eastward over Niagara Falls and the still-active industrial zone of Ontario, past the steel mills of Hamilton and the urban density of Toronto, and eventually winds up in the Saint Laurence Seaway and then the Atlantic Ocean. Surrounded by a fossil fuel capitalist density of refining, steel, and car manufacture, the Great Lakes are the largest basin of fresh surface water on the planet, holding some 84 percent of North America's fresh surface water (US Environmental Protection Agency). The Great Lakes are Indigenous lands, with Lake Ontario, in Anishinaabe and Haudenosaunee territories, severed in half by two settler nation-states, the United States and Canada. The water of the lakes is in movement; it is indifferent to borders; it cycles through the atmosphere, splashes off office towers, and returns to the water that people, plants, and animals all rely on. We, humans and nonhuman beings, the lake and the city, the settler colonial industry and Indigenous sovereignties, are all dependent on the water's many forms and powers. More than this, we are part of the being of the water; we are bodies of water. Water does not just support life, Anishinaabe Elders and water keepers teach: it is life; it is alive (McGregor 2009). Water is life. And people are made by, contain, and are beholden to this greater liveliness.

And, since the mid-twentieth century, we have been made by polychlorinated biphenyls too. After analyzing urine, blood, and breast milk, twenty-first-century global biomonitoring studies have concluded that all people alive today contain PCBs within them, along with DDE, the metabolite of DDT (United Nations Environmental Programme 2009). Industrially produced chemicals such as PCBs have become a part of human living-being. Conventional sciences treat these chemicals as the elementary building block of the world, as merely atoms forming molecules that add up to our substance. But such an atomized understanding of industrial chemicals leaves out their extensive relationships and enactments of structural violence, as well as the ways that they have become part of living-being in a pollution-saturated world. Taking a decolonial approach to industrial chemicals is crucial to reimagining the elemental as an unevenly distributed, responsibility-filled, and constituting domain binding water, air, violence, molecules, living beings, and futures together in pain and potential. This chapter therefore starts with the ways ongoing pollution under colonialism joins the molecular fabric of our bodies and waters and then moves to a speculative rethinking of what chemicals are. Moreover, this chapter's attention to chemical pollutants and the project of reimagining elements struggles with the ways it is epistemically caught in the work of disciplines such as toxicology, epigenomics, environmental chemistry, and so on, as well as decades of environmental-justice community knowledge making. Mindful of this multiplicity of noninnocent modes of knowing chemicals, this chapter attempts an anticolonial study of industrial chemicals, starting with PCBs, and in doing so tries to reimagine chemicals with and against technoscience.

If we agree that dominant modes of technoscience also enact infrastructures of colonialism, capitalism, militarism, and white supremacy, then it might seem that decolonizing technoscience is an impossibility. Many pedagogies of science and technology studies (STS), at least as they are used in the United States, Canada, and Europe, still focus on the accomplishments of mostly white men in expensive labs in white spaces and ignore the abundance of decolonial tactics already in the world. This decolonial project of studying industrial chemicals with and against technoscience starts with an alternative genealogy for STS, one which hails from Frantz Fanon and other radical Black, Indigenous, feminist, and queer thinkers. I recognize that my decolonial politics are from a particular place—urban Tkaronto—where particular local and diasporic political and intellectual itineraries meet. In Tkaronto, decolonial horizons are toward Indigenous sovereignty that centrally include Indigenous governance of lands, waters, and territories. At the same time, Tkaronto is a place where many diasporic decolonial politics converge, where projects of Black emancipation join with queer liberation to meet Indigenous resurgence on stolen land. In Tkaronto, a decolonial horizon is necessarily aimed at refusing the settler state. Thus, the decolonial tactics offered up here are not meant as a universal recipe but rather are done with humility, with the awareness that other knowledge itineraries meet elsewhere that I have much to learn from.

Learning from Fanon (1963, 1965, 1967) as one starting point for a decolonial STS, we might attend to how he, as a Black doctor from Martinique who trained in Paris and lived in Algeria, struggled within the contradictions that make up both science and racial embodiments. We might notice how he theorized within a set of contradictions—for example, between his denouncement of the ways Blackness is dehumanized in colonial and racist worlds and his embrace of Blackness toward a new humanity, or his rejection of medicine's racist pathologizing of colonial suffering and his call for self-directed medical care as part of decolonial struggle. Fanon rejected the violence of colonial medicine but tried to remake it anyway. He sought to dismantle and to reinvent.

Learning from and making kin with his decolonial projects, how might we take up the anticolonial call to dismantle and also embrace a project of inserting an otherwise into the world, following Fanon's (1963, 229) sense that "the real leap consists in introducing invention into existence"? What has to be seized, rebuilt, and dismantled to make room for decolonial potentials? Private property? White supremacy? Fossil fuel economies? Militarized policing? Extractive research? There are so many structures to refuse. In following this path, we can also take direction from practices and theorizations of Indigenous refusal, which are not only central to Indigenous feminist studies (as exemplified by Audra Simpson and Eve Tuck, among others) but also pivotal to any decolonial feminist politics of consent—where consent is not a practice of getting to a contractual *yes* but instead an invitation to the *no* in ongoing commitments to permission, reciprocity, and relationship making (Audra Simpson 2014; Tuck and Yang 2013). Pollutants are fundamentally about consent: they are nonconsensual disruptions and harms to lands, Indigenous sovereignty, and bodies—disruptions that are made possible by the colonial state's racist claim to the right to give permission to enact violence on particular places, beings, and peoples. Indigenous feminism says *no* to both this nonconsensual violence and the mode of colonial governance through which the colonial state allows pollution as a legitimate activity.

In these particular ways, this chapter is about refusing and dismantling, while also inserting invention into existence and making decolonial futures in relation to industrial chemicals, with and against technoscience. Industrial chemicals, as given to us by engineering, fossil fuel companies, Big Pharma, and capitalism, will not do. Our inherited models of the chemical, I believe, need undoing and rebirth, if we are to stay in the tension of refusing violence and becoming something else.

So how do we dismantle our sense of the industrial chemical? I begin with the refusal of three epistemic habits that run through almost all of the technoscientific work materializing chemical violence: portraying chemicals as merely molecular elements, obscuring the violence of chemicals, and damage-based research. From there, the chapter will move to the task of inserting invention into existence and end by speculatively reimagining how industrial chemicals might be remodeled toward decolonial horizons of responsibility and better worlds.

REFUSING THREE COLONIAL EPISTEMIC HABITS

Industrial chemicals are largely made known to us through the knowledge practices of the very entities—engineering, corporations, the military, and the state—whose violence we seek to dismantle. To begin to say *no* to this version of the chemical, involves refusing the portrayal of chemicals as discrete entities, often modeled through abstract structural diagrams that itemize singular molecules as an arrangement of atoms. In such representations, chemicals float in white space, largely without relation. All such diagrams are only *models* of chemicals, not the things in themselves. The structural chemical model in particular was created with engineering and industrial needs in mind and as a tool to create a taxonomy of substances that could be heralded as proprietary inventions (Hepler-Smith 2015). This dominant model of the chemical as molecule blocks our attention from origins, histories, profits, and violence. It scales the chemical as small and invisible to the eye, rather than large, extensive, and permeating. This structural representation of discrete molecules is built into the ways toxicology has historically studied chemicals: one by one, as isolated entities of purely technical qualities without context. The extensive cloud of synthetic chemical relations, whether emitting from factories, extraction, agriculture, infrastructures, or commodities, is erased by this model. Externalizing the complexity of chemicals works well for capitalist ledger books that structurally will not count side effects, fallouts, and discards. Today, this way of modeling chemicals as molecules seems ordinary and self-evident (Myers 2015). It is difficult to talk about chemicals in any other way, regardless of your politics. As a result, the expansive scope of chemical relations that surround and make us largely resides in the realm of the imperceptible. We might feel some of our chemical relations, and we might even intensely experience the pain they cause, but the fullness of our chemical relations ends up being largely conjectural.

The second epistemic habit I want to refuse builds on the first. In this habit, infrastructures of environmental monitoring and research are, for the most part, projects meant to obscure chemical violence; they are purposefully built infrastructures of not knowing. State-collected environmental data can rarely be trusted. It is not only that chemical pollution itself is a material kind of colonial and racial capitalist violence disrupting life and land; it is also that the very infrastructures governments build to detect and care for our chemical surround are themselves a form of colonialism. The regulatory systems that grant permission to pollute are designed to deny violence, even as that violence occurs in plain sight. Building on Max Liboiron's (2021) work that tracks the ways in which pollution is a form of colonialism, this chapter emphasizes that state regulatory systems are largely permission-to-pollute infrastructures that enact colonialism. In the name of regulating pollution, settler colonial governance is designed to allow pollution. Government regulations legalize pollution up to a certain level or even beyond it, as long as the pollution is reported in some way. Permission-to-pollute state systems both facilitate and erase environmental violence and thus are also part of colonialism, and this too can be refused. Thus, the equation that the pamphlet *Pollution Is Colonialism* (written by Max Liboiron, Dayna Nadine Scott, Natasha Myers, Reena Shadaan, and Michelle Murphy) offers is "Pollution + Permission-to-Pollute-State = Colonialism" (Endocrine Disruptors Action Group and CLEAR 2017).

There is a thick archive of examples of this equation, and this chapter takes up one such example of its manifestation in setter colonial worlds. On February 23, 2017, flames rose far above the stacks of the Imperial Oil Refinery in Sarnia, Ontario's "Chemical Valley" that runs along the Saint Clair River, a shipping channel that connects Lake Huron with Lake Erie and also serves as a border between Canada and the United States. Standing on the US side of the river, community members used their smartphones to capture videos of the enormous flames surrounding the refinery and licking upward (CTV News Windsor 2017).

Chemical Valley is the colloquial name for the Sarnia Lambton industrial corridor responsible for over 40 percent of Canada's petrochemical processing. Imperial Oil began in the 1880s in the Sarnia area, a region with the world's first commercial oil patch (founded in 1858) which was then dubbed Canada's Oil Lands. The area known as Chemical Valley was built on land stolen from Anishinaabeg people in 1919, when an Indian Affairs agent coerced the community of the Aamjiwnanng First Nation (then called Sarnia Reserve 45) into selling 1,184 acres under the threat of federal expropriation of all their lands. The sale price was never paid.

Imperial Oil, now owned by Exxon-Mobil (previously called Standard Oil), built its more industrialized refinery in 1920. When World War II broke out, the infrastructures of oil refining laid the groundwork for establishing a national center of chemical production, hitched up through shipping lanes, railroads, and later pipelines across the continent and to the planetary scale of the multinational conglomerate of Exxon-Mobil. Aamjiwnanng First Nation is now surrounded by over fifty petrochemical plants (MacDonald and Rang 2007).

The Imperial Oil refinery today celebrates itself as a state-of-the-art oil and chemical processing and research facility, processing some 120,000 barrels of crude oil a day. In addition to oil, the plant makes polyethylene, solvents, higher olefins, benzene, polyurethane, xylene, and intermediary chemicals that are then used to make detergents, lubricants, hydraulic fluids, and drilling fluids.

So what are you watching when you see this cellphone video image of flames engulfing the facility and turning the sky orange? Figure 11.1 shows Imperial Oil's response. In essence, the message was: Grassfire put out, no air quality issues detected, no injuries. Nothing happened here. So, what were we watching? It is likely that readers will find this kind of corporate double-talk familiar. At the same time, perhaps so is the sensation of doubting your own senses. What just happened? Am I sure I know what I am experiencing? Imperial Oil is required by law to release only a tweet-size report to the public through the municipal, industry-organized Chemical Valley Emergency Coordinating Organization (CVECO) and later sends self-reported information to the provincial ministry of environment. In Chemical Valley, CVECO reports come in a steady stream (Luginaah, Smith, and Lockridge 2010; Plain 2016). Meanwhile, Imperial Oil's fence-line monitor is set up to detect only six chemicals. Together, this narrowly designed monitoring and reporting system performs a type of erasure of all the other emissions that are not surveilled. It allows Imperial Oil to claim that no emissions were detected. The local industry association, rather than the government, does all the air monitoring in the petrochemical refining corridor. Virtually all monitoring consists of industry self-reporting. Thus, the primary function of the monitoring, which is actually a nonmonitoring, is precisely to perform an erasure of environmental violence.

I think of this practice of monitoring as erasure as a kind of *gaslighting*. I only recently learned about the term from a viral opinion piece by the impressive feminist reporters at *Teen Vogue* (Lauren Duca, "Donald

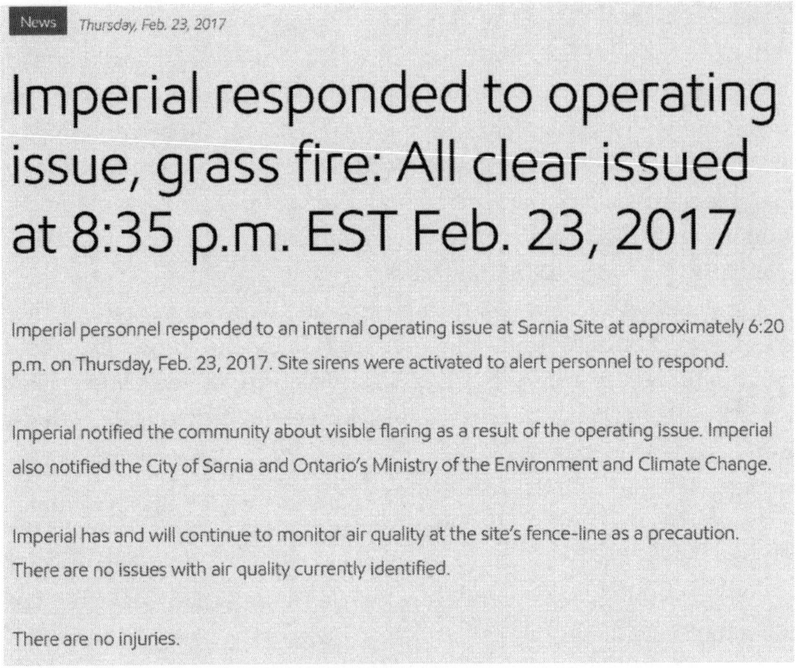

Imperial responded to operating issue, grass fire: All clear issued at 8:35 p.m. EST Feb. 23, 2017

Imperial personnel responded to an internal operating issue at Sarnia Site at approximately 6:20 p.m. on Thursday, Feb. 23, 2017. Site sirens were activated to alert personnel to respond.

Imperial notified the community about visible flaring as a result of the operating issue. Imperial also notified the City of Sarnia and Ontario's Ministry of the Environment and Climate Change.

Imperial has and will continue to monitor air quality at the site's fence-line as a precaution. There are no issues with air quality currently identified.

There are no injuries.

FIGURE 11.1: Imperial Oil Sarnia Refinery's public statement released on its corporate website about the large flaring event on February 23, 2017. This information is no longer posted on its website. https://news.imperialoil.ca/press-release /imperial-responded-operating-issue-grass-fire-all-clear-issued-835-pm-est-feb -23-2017. Accessed May 11, 2017.

Trump Is Gaslighting America," December 10, 2016). *Gaslighting* is used by psychologists and among domestic violence support communities to name a kind of psychic interpersonal abuse. The term comes from a 1938 play, which was made into a 1944 movie starring Ingrid Bergman, about a woman whose husband makes her question her own sanity and trust of her senses. She even wonders whether she really saw the flickering of the gas-fueled lighting in her home, a dimming which would indicate that someone was inhabiting the house when the husband claimed to be away (Cukor 1944; Hamilton 1939). Gaslighting works by a variety of techniques, including blatantly lying, denying events took place even when there is proof, hiding things, moving the goalposts so you can never get it right, deflecting from the subject at hand, counteraccusing the victim of lying and manipulating, and entangling compliments and statements of care with abuse.

Gaslighting is not just an interpersonal phenomenon. It is also infrastructural—where ways of not knowing about chemical exposures, their itineraries, and their pervasive presence have been built into our world. The invisibility of our chemical surround is not a natural fact; it is an achievement. In other words, the invisibility of chemicals is due not only to the uncertain properties they have but, more profoundly, to concerted state and corporate efforts to prevent their technological perception. Corporate and state gaslighting is pervasive and bold. We experience it every day. We experience it on our screens and Twitter threads, in our communities and homes, in our bodies and feelings. Gaslighting is nothing new, even as it is pervasive at this moment.

Industry self-reported environmental data are a key example of what I call infrastructures of gaslighting. This phenomenon builds on a variety of efforts to produce doubt by industrial lobbies, agnotology, and regimes of imperceptibility (Brandt 2007; Markowitz and Rosner 2003; M. Murphy 2006; Oreskes and Conway 2010; Proctor and Schiebinger 2008). Gaslighting as a concept highlights reality distortion, rather than the production of uncertainty, and emphasizes its psychic effects. Thus the concept also builds on work by Fanon (1967), Du Bois (1903), and others on the abusive psychic effects of white supremacy's denying hostile worlds. At the same time, it connects the erasure of environmental violence to forms of gendered abuse and thus seeks to contribute to work in Indigenous feminism that highlights the entanglements between extraction, environmental violence, and sexual violence (Native Youth Sexual Health Network and Women's Earth Alliance 2016). Lastly, the sense of infrastructures of gaslighting emphasizes the government as a key actor in racial capitalism, white supremacy, environmental violence, and settler colonialism (Pulido 2017).

Exxon-Mobil, which owns Imperial Oil, is an infamous "merchant of doubt" corporation which spent three decades developing a campaign to deny climate change (Supran and Oreskes 2017). Gaslighting is not just interpersonal; it is part of intentionally built colonial, racial, and capitalist practices used to deny the violence of their structures. In other words, the purposeful creation of the surround of violent externalities depends on infrastructures built out of techniques of gaslighting.

From the departure point of infrastructures of gaslighting, we can begin to ask a set questions: How do these infrastructures work? What practices are they made of? How do they fail? What irrealities make up our surround? What genealogies converge to make up the infrastructures of gaslighting? Infrastructures of gaslighting are not just or especially environmental. Many, perhaps most, people live daily in conditions that deny the materiality of their experience. Gaslighting is foundational to how settler colonialism works. The legal concept of terra nullius (land as empty) that undergirds the settler colonialism doctrine of discovery, which in turn makes possible Canada's claim to sovereignty, is based on the denial of existing Indigenous presence. This is exemplified in the very building of Chemical Valley on stolen land. As Patrick Wolfe and many other scholars have argued, settler colonialism fundamentally operates through a logic of Indigenous elimination and erasure (Culthard 2014; Kauanui and Wolfe 2012; Wolfe 2006). Moreover, gaslighting is also part of how whiteness works, including more subtle ways, such as when all-white work spaces are portrayed as collegial, nice, and pleasant by the white people who inhabit them, denying the unwelcoming environment they create for people of color. White supremacy at its most destructive is a gaslighting project as well, wrapping itself in color-blind ideologies such as All Lives Matter to deflect from its commitment to racist violence. Everyday infrastructures of gaslighting disrupt people's ability to lay claim to their own account of reality without psychic fracturing, but they are also materially deadly (Davis and Ernst 2019).

In Chemical Valley—where there is a cluster of over fifty petrochemical plants—gaslighting is a constant part of life, reflected in the regular sirens and steady public announcements of spills and accidents while little information is provided and communities are told there is nothing to worry about (MacDonald and Rang 2007). Compared to the detailed and intrusive surveillance of life that corporations and the state are able to conduct in the age of social media and big data, the overwhelming lack of monitoring of chemical violence is striking. For at least a century, there has been a struggle with corporate gaslighting techniques about the reality of injurious, human-made chemical exposures in bodies, workplaces, communities, and environments.

Since Imperial Oil cannot be depended on to provide the evidence of its own violence, one might then look up the company's emissions data in Canada's mandatory National Pollution Release Inventory (NPRI). These data, which are reported annually by factories, mines, refineries, extraction sites, and other large enterprises, are mapped and posted online and can then be used by environmental justice organizations to show environmental racism and the concentrated exposure to pollution in Indigenous territories.

Look a bit closer, though, and you will see that the emissions data are industry self-reported. Moreover, they tend not to be based on any physical measure of emissions at all. Rather, much of the reported data is purely an estimate based on a kind of calculation that relies on standardized "emission factors." These factors, in turn, are also estimates—an estimated "average" emission level of a substance produced by a particular kind of industrial activity. The emission factors are published by the Environmental Protection Agency (in a technical document called AP-42 which goes back to 1972 and is only irregularly updated), where they are compiled based on existing reports and research that vary tremendously in quality and are largely generated by the industries themselves. It is well known that emissions factors can be based on old and inadequate research and often rely only on estimates and not actual physical measures of pollution effects. Therefore, not only are the NPRI data about emissions not grounded in physical measures they are often a quantitative fiction based on further layers of historical industry self-reporting. The data themselves are a kind of deflection, a performance that hides a longer chain of corporate efforts to avoid having their activities measured. The data offer the pretense of state monitoring, erasing the fact of nonmonitoring. State environmental data concerning chemical exposures are thus designed to perform this twisted psychic abuse, purposely built to deflect attention and wrap falsehoods in the veneer of state care. Chemical pollution is thus reflected back to you in state-collected data disguised as efficiencies and stewardship. Disasters are not happening, and you are not being killed, even as you can feel it happening. Are we to feel lucky for these abusive accountings, because the alternative is to have no data at all? We are caught in worlds made to deny the terms of our very existence.

The third technoscientific habit shaping industrial chemicals that I want to refuse follows the first two: because it is so rare for our chemical surround to be documented, technoscientific research that seeks to contest the presence of synthetic chemicals in the world tends to proceed by detecting and measuring the damage chemicals do to bodies and lands. (So, too, do ethnography and the social sciences more broadly often seek to bear witness to the evidence of damage.) This type of research (often biomedical) puts the burden of representing violence on already dispossessed communities. It is hard to perceive the infrastructures of chemical violence in the world at the same time that there is an abundance of research attending in exquisite detail to the pathological molecular metabolisms of chemicals in bodies and the symptomologies in communities already living in hostile conditions. This damage-based research has pernicious effects, bringing focus to chemical violence by virtue of rendering lives and lands as pathological, as damaged and doomed, as inhabiting irreparable states that are not just unwanted but less than fully human.

Refusing this economy of research, Indigenous feminist Unangax̂ scholar Eve Tuck has called for "suspending damage" as a refusal to participate in "damage-based research" (Tuck 2009; Tuck and Yang 2013). This refusal is a challenge to environmental justice habits. It is an invitation to find other ways to shine critical light on the infernal entanglements of colonial capitalism as expressed through chemical relations, and at the same time it is a call to direct creative energy toward decolonial possibilities.

How does one refuse these three habits: the chemical as an isolated entity, the gaslighting of permission-to-pollute data, and damage-based research? How can one be with and against technoscience and not only refuse and dismantle but offer alternative objects for each other and toward decolonial futures? Decolonial STS cannot just critique technoscience; it must offer other concepts and practices that remake it.

STEPPING INTO SOMETHING ELSE

What would an alternate sense of the chemical look like if it were made to affirm Indigenous sovereignty, hold perpetrators responsible, and love life and land even in their violated forms?

The resurgent practices of the Native Youth Sexual Health Network in Toronto have created powerful activations of Indigenous reproductive and environmental justice that guide me. A fundamental value in their work is that "violence on the land is violence on our bodies" (Native Youth Sexual Health Network and Women's Earth Alliance 2016). We are always on and of the land. Our ordinary acts of transporting, purchasing, eating, and sheltering are caught up on long arcs and wide networks of extractive colonialism. Being in a city, or next to a refinery, is still being in body-land relation. What would an alternate mode of the chemical look like that holds violent systems responsible for their disruptions to land-body relations that are themselves honored?

I have been trying to both refuse these three epistemic habits for studying chemicals and step into something else during my research into the history of PCBs on the lower Great Lakes. Polychlorinated biphenyls are the classic persistent chemical and one of the few chemicals that are banned. Their manufacture in North America was banned in the late 1970s and globally through the Stockholm convention in 2004, and yet they have spread into the airs, waters, soils, and foodways of the entire planet. The term *PCBs* is plural, as there are 209 different ways of attaching chlorines to biphenyls and each is a little different, so the category hides many important distinctions of relation. In general, PCBs tend to be quite inert, resistant to temperature change, and insoluble in water. They are slow to break down, to biomagnify in the food chain, and to concentrate in fatty tissues. Almost all PCBs were manufactured by one company, Monsanto, which also brought the world DDT, Agent Orange, and the glyphosate-based pesticide Round Up. After the industrial exuberance is gone and Monsanto moves on to the next unregulated chemical product, the PCBs remain. Polychlorinated biphenyls are in every person alive and have been found in high concentrations in the deepest trenches of the ocean, as well as in the Arctic. Despite being banned, PCBs are still used. While their *manufacture* was banned, PCB *use* has been "grandparented": older installations are allowed to persist if they meet certain conditions of containment. They are thus still part of the aging electrical grid of the Great Lakes. When people in Toronto switch on a light, they are entangled in PCBs' expansive relations. Polychlorinated biphenyls are in the power grid that lights up cities, in the transformers that carry electricity, and in the electrical wiring in homes. We are caught in their circulation daily. At the same time, it is crucial not to forget that concentrated exposure to PCBs in the Great Lakes area is the legacy of unregulated industry run by the logics of racial capitalism built on stolen Indigenous land that purposely pushed, buried, and dumped its chemical violence on Indigenous, Black, and poor communities. Polychlorinated biphenyls are a rare example of a highly regulated and monitored industrial chemical that has nonetheless comprehensively infiltrated the planet. Yet, before they were regulated, they created concentrated racialized violence. Civil rights organizing in 1982 against the dumping of PCBs in the Black-majority community of Warren County, North Carolina, in the United States gave us the term *environmental racism* (McGurty 2007; United Church of Christ Commission on Racial Justice 1987). Polychlorinated biphenyls teach us that how we do our chemical relations today extends colonialism and racism far into the future. Chemical exposures have been purposefully built into our worlds. We are in futures already altered, as well as in the lag between past exposure and symptom.

Polychlorinated biphenyls are not only persistent; they are also endocrine disrupting. Their shape is similar to that of the human thyroid hormone, which makes it easier for PCBs to participate in human metabolism, including the regulation of gene expression, which in turn rearranges how metabolism works. Moreover, there is strong evidence to suggest that PCBs can be part of a kind of epigenetic inheritance, in which their metabolic effects carry over across generations. Thus, PCBs have become part of our chemical relations, not just because they are persistent in lands and waters but because they are part of our metabolic intergenerational inheritance. Vanessa Agard-Jones (2013) describes a disturbing kind of "chemical kinship" that synthetic chemicals can create across bodies. This chemical kinship is not just about solidarity in oppression; relatives are not necessarily chosen and can nonconsensually bind us to violent relatives of plantation owners, chemical companies, and settler states. This makes Monsanto, which manufactured most of the world's PCBs, a grandfather, or grand-kin of sorts, to all humans of the early twentieth century (for the history of environmental justice around the Alliston, Alabama, chemical plant that first manufactured PCBs, see Spears 2016). How to hold these bad kin responsible, who have forcibly intruded into our bodies, development, relations, and becoming?

The history of the spread of banned PCBs and the pervasiveness of environmental gaslighting suggest that the ways chemicals have been monitored by states have failed to prevent their violence. I want a reinvention of the chemical that is unflinchingly pessimistic in its acknowledgment that these chemical relations are racist, harmful, and even deadly and that will look for ways to hold us all accountable for being caught up in killing, even if these chemicals are killing us too, just more softly.[1]

This work reconceiving chemical relations has largely happened while I have been living in Toronto on Anishinaabe territory, as a Métis person of white skin, whose father is white and mother Métis. I have done this work as an urban white Métis professor with many uneasy responsibilities and as someone who has been structurally complicit in erasures and colonial institutions such as the university. I do this work in a settler colonial world that is continually resisted by Indigenous land defense and water protection. I confess that my approach to refusing the colonial logics of chemical exposures is a project of trying to figure out responsibilities in and to the place where I am a guest. So, with and against technoscience, I want to ask how to build a speculative, decolonial model of chemical violence and how to insert invention into existence (Fanon 1963) while also dismantling the colonial logic of erasure built into so much government environmental monitoring.

In reimagining the chemical, I am inspired by the art of the Métis artist and Indigenous reproductive justice defender Erin Marie Konsmo. The report *Violence on the Land, Violence on Our Bodies* (Native Youth Sexual Health Network and Women's Earth Alliance 2016) includes a set of stencils that portray lungs holding within them the violent infrastructure of extractive colonialism. The stencils are labeled "Violence from Fracking Is Violence on Our Bodies," "Violence from Pipelines Is Violence on Our Bodies," and "Violence from Logging Is Violence on Our Bodies." These stencils are a starting point for revisualizing the scale and scope of chemical relations (see Michelle Murphy 2017; for images, see Women's Earth Alliance and Native Youth Sexual Health Network 2016).

The campaign invites people to hold the images over their chest and take a picture to post on social media (using the hashtags #landbodydefense and #environmentalviolence)—resulting in poses of Indigenous women, two-spirit people, and youth containing the extensive infrastructural relations of settler colonial capitalism inside them. The land is inside us. Both Indigenous people and settlers have responsibilities to figure out. Life forged in ongoing chemical violence is also life open to becoming something else still, which is not a nostalgic return but instead the defending of Indigenous sovereignty and land, starting here, within oneself and each other, here in the damage now. There is no waiting for the better condition to arrive.

The stenciled portrayal of pipelines inside lungs subvert a normative sense of scale, where the small nests in the large; where tiny chemicals enter bodies surrounded by larger infrastructures. In this normative sense of scale, the industrial chemical is rendered only as small, even as it is pervasive or as the extension of a system. The stencils offer another sense of land-body relations, in which the expansive relations of infrastructures exist inside bodies.

Why must our sense of chemicals stay small? How might an altermodel of the chemical manifest its expansive relations, out into land-body relations, settler colonialism, white supremacy, infrastructures, capitalist relations, heteronormativities, nation-states, kinships? I believe we need a version of the chemical that rejects the separation of knowledges that

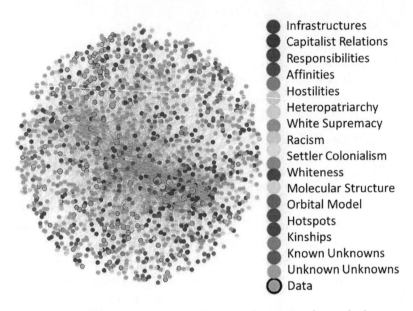

Infrastructures
Capitalist Relations
Responsibilities
Affinities
Hostilities
Heteropatriarchy
White Supremacy
Racism
Settler Colonialism
Whiteness
Molecular Structure
Orbital Model
Hotspots
Kinships
Known Unknowns
Unknown Unknowns
Data

FIGURE 11.2: Speculative cut-and-paste diagram of expansive chemical relations. The relations imagined here are ethical and not merely abstract or physical. Image by author.

kept molecules as a concern only of the technosciences, that confined chemical relations to the atomic. I want to refuse the model of the molecule hovering in white space. An altermodel of chemicals, such as the one imagined in figure 11.2, would manifest extensive relations, honoring land-body relations, drawing on a multitude of knowledges, toward a sense of tracing and activating the density of responsibilities and ways we make and interrupt each other.

This is an invitation to imagine such an extensive model of the chemical. When you trace an industrial chemical in a body, how might we invite one another to attend to the extensive relations that become part of us? We are more than individuals. If we can think of bodies as already multispecies beings, as collectivities and clouds of organisms, from microbes to humans, can we also imagine along with figure 11.3 how living-being is also constituted through these extensive chemical relations that are part of our being?

If the evolution of life is understood through relational symbiogenesis processes that, for example, evolved eukaryotic life out of the partial absorption of single-celled beings into one another, then our altermodel of chemicals might upend the thinking about what constitutes life in the

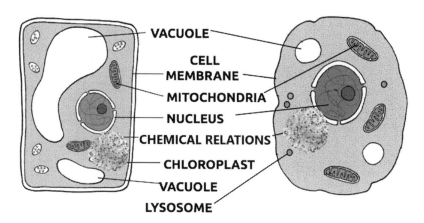

FIGURE 11.3: Imagining the renarrating of symbiogenesis to include the non-consensual participation of industrial chemicals in the development, gene regulation, and epigenetic relations of life. Here, instead of imagining industrial chemicals as a molecule entering the body and interacting at the biomolecular scale of a receptor, for example, industrial chemicals are modeled as carrying with them their extensive relations, including with colonialism. Image by author.

age of ubiquitous presence of isotopes from nuclear testing, persistent organic pollutants, and endocrine-disrupting chemicals as constituent elements of living-being (Margulis and Sagan 2000). How are chemicals part of a violence-infused symbiogenesis?

The very understanding of what makes up our molecular inheritance (usually confined to DNA birthright) is called into new questions by acknowledging the industrial chemicals that connect us and participate in our biological development. These chemical relations are, moreover, mostly nonconsensual. We are caught in them, even if we participate in their generation as the condition of our living. They are forcibly inserted into the world. They are disruptions into community orders and lifeways; they interrupt bodily and Indigenous sovereignties. If this is so, then, let us pull our attention to these chemical relations back outward to the infrastructural register, back to the corporation and the state, and point out responsibilities for inheritance and disruption back at Monsanto. Instead of burdening injured bodies and lands with the work of representing this violence (as is the norm in damage-based research), can we make our chemical models point the violence back at perpetrators, not inside bodies? Figure 11.4 attempts to visualize this sense of nested infrastructural responsibility.

FIGURE 11.4: *Attaching Responsibility to Injurious Kin.* This speculative image attempts to imagine placing the representational burden of violence on the corporate perpetrator of pollution. Here, the example is of Monsanto, which manufactured almost all of the world's PCBs, as well as DDT and Agent Orange. Image by author.

One of the largest stockpiles of PCBs on the northern side of Lake Ontario is in downtown Tkaronto/Toronto, in the high-rises of the financial center. Canadian bank towers not only collect PCBs; they actively use them in their electricity grids that power their servers and lights. Finance capital is saturated with PCBs. More than this, Canadian banks are central players in the histories and infrastructures of extraction. It is worth remembering that Canada was formed as an extraction nation. The Hudson Bay Company passed its doctrine of discovery charter to the new nation of Canada in 1867. Toronto-Dominion Bank, known as TD Bank, is among Canada's oldest banks, founded in 1871, and today is the nineteenth biggest bank in the world, as well as a major funder of extraction and pipelines in North America. Its securities division regularly promotes pipelines within its portfolio as fabulous investments, with secure rates of return. When the pipelines make no sense to communities, futures, environments, or even national economies, they nonetheless make sense to finance capital, and it is finance capital speculation of stockholder-owned multinationals that continues to propel fossil fuel extraction in North America. Toronto-Dominion and extraction have long been entangled. Dominion Bank was founded in 1871 by the Austins of Toronto, the same family that founded Consumer Gas in 1848, which would later become Enbridge, which was initially incorporated in 1949 by Imperial Oil. Enbridge has the world's biggest extension of oil pipelines, with 30,040 kilometers, as well as holding partial shares in many other pipeline projects. The divestment campaign supporting the struggle against the Dakota Access Pipeline by the Standing Rock Sioux explicitly targeted TD Bank as a crucial funder. The bank, Enbridge, and Imperial Oil are in a close kinship relation.

Thus, in the effort to reimagine the extensive relations of industrial chemicals, might we attach those relations not only to Monsanto as a manufacturer of chemicals or to Imperial Oil as a point source polluter, but also to TD Bank, and hence to finance capital? "Violence from Finance Capital is Violence on Our Bodies": Imagine a stencil with the extensive infrastructure of finance capital inside your lungs. Imagine refusing the current state-industry environmental data systems and their gaslighting practices and instead building other, more just kinds of environmental data practices. Can we attach those lungs, and the extensive relations of chemicals, back onto TD Bank, back to the settler state, to support new forms of refusal and responsibility? This is an invitation to build alternative environmental data justice infrastructures that seek to dismantle and reinvent. While this speculative vision of an altermodel of chemical violence offered in figures 11.5 and 11.6 are made with crude, scale-busting, cut-and-paste skills, building a prototype of such a model is possible.

I invite you back to that rainy, warm winter night in Tkaronto and ask you to breathe with me. Breathe in. With each inhalation, the expansive molecular relations of corporate kin are pulled into your lungs and attach to receptors, triggering metabolic shifts, altering gene expression,

**Violence From Finance Capital
Is Violence On Our Bodies**
#EnvironmentalViolence

FIGURE 11.5: Imagining an additional stencil that adds finance capital to the nested scaling of responsibility for violence on land and bodies. Finance capital props up companies such as Imperial Oil, Enbridge, and Monsanto and is a further layer of responsibility for environmental violence. Additionally we could imagine a stencil with the colonial state or gaslighting data practices inside our lungs. Image by author. Original stencil by Metis Indigenous feminist land and body defender Erin Marie Konsmo as part of the action toolkit for the launch of *Violence on the Land, Violence on Our Bodies* (Women's Earth Alliance and Native Youth Sexual Health Network 2016).

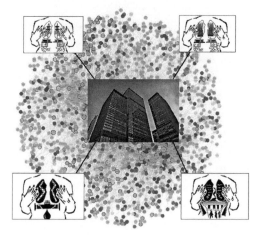

FIGURE 11.6: This image speculatively models another way of attaching responsibility for damage to finance capital by building it right into our ways for modeling what chemicals are, rather than as damage done to bodies. These simplistic cut-and-paste images attempt to gesture toward alternative ways of understanding the extension and nested scales of what chemicals are in relationship to bodies, lands, and perpetrators. Image by author.

becoming us. Breathe out. Breathe in. With each exhalation, feel the fragility of infrastructures of gaslighting around you. Breathe out. Once more, breathe in. Feel yourself reconnecting to the greater fulsomeness of our relations, to the possibility of becoming otherwise, to land and all our relations, to another world of consent and care. Breathe out. Breathe in with and against the science and toward something else, not waiting for a better moment to arrive.

NOTES

This chapter expands on Murphy 2017.
1. This phrasing is inspired by Harney and Moten 2013, 10.

REFERENCES

Agard-Jones, Vanessa. 2013. "Bodies in the System." *Small Axe* 17, no. 3 (42): 182–92.

Brandt, Allan. 2007. *The Cigarette Century: The Rise, Fall, and Deadly Persistence of the Product That Defined America*. New York: Basic Books.

CTV News Windsor. 2017. "Operating Issue Causes Flaring at Imperial Oil," February 24. https://windsor.ctvnews.ca/operating-issue-causes-flaring-at-imperial-oil-1.3299818.

Cukor, George, dir. 1944. *Gaslight*. Culver City, CA: Metro-Goldwyn-Mayer.

Culthard, Glen. 2014. *Red Skin, White Masks: Rejecting the Colonial Politics of Recognition*. Minneapolis: University of Minnesota Press.

Davis, Angelique M., and Rose Ernst. 2019. "Racial Gaslighting." *Politics, Groups, and Identities* 7 (4): 761–74. https://doi.org/10.1080/21565503.2017.1403934.

Du Bois, W. E. B. 1903. *The Souls of Black Folk*. Chicago: A. C. McClurg.

Endocrine Disruptors Action Group and CLEAR. 2017. *Pollution Is Colonialism*. Toronto: EDAction and CLEAR.

Fanon, Frantz. 1963. *The Wretched of the Earth*. Translated by Constance Farrington. New York: Grove Press.

Fanon, Frantz. 1965. "Medicine and Colonialism." In *A Dying Colonialism*. Translated by Haakon Chevalier, 121–46. New York: Grove.

Fanon, Frantz. 1967. *Black Skin, White Masks*. Translated by Charles Lam Markmann. New York: Grove.

Hamilton, Patrick. 1939. *Gas Light: A Victorian Thriller in Three Acts*. London: Constable.

Harney, Stefano, and Fred Moten. 2013. *The Undercommons: Fugitive Planning and Black Study*. New York: Autonomedia.

Hepler-Smith, Evan. 2015. "'Just as the Structural Formula Does': Names, Diagrams, and the Structure of Organic Chemistry at the 1892 Geneva Nomen-

clature Congress." *Ambix* 62 (1): 1–28. https://doi.org/10.1179/1745823414Y
.0000000006.

Kauanui, J. Kēhaulani, and Patrick Wolfe. 2012. "Settler Colonialism Then and
Now: A Conversation Between." *Politica and Società* 2:235–58. https://doi.org
/10.4476/37055.

Liboiron, Max. 2021. *Pollution Is Colonialism.* Durham, NC: Duke University Press.

Luginaah, I., K. Smith, and A. Lockridge. 2010. "Surrounded by Chemical Valley
and 'Living in a Bubble': The Case of the Aamjiwnaang First Nation, Ontario."
Journal of Environmental Planning and Management 53 (3): 353–70.

MacDonald, Elaine, and Sarah Rang. 2007. *Exposing Canada's Chemical Valley: An
Investigation of Cumulative Air Pollution Emission in the Sarnia, Ontario Area.*
Toronto: EcoJustice.

Margulis, Lynn, and Dorion Sagan. 2000. *What Is Life?* Berkeley: University of
California Press.

Markowitz, G., and D. Rosner. 2003. *Deceit and Denial: The Deadly Politics of Industrial Pollution.* Berkeley: University of California Press.

McGregor, Deborah. 2009. "Honouring All Relations: An Anishnaabe Perspective
on Environmental Justice." In *Speaking for Ourselves: Environmental Justice in
Canada,* edited by J. Agyeman, R. Haluza-Delay, P. O'Riley, and P. Cole, 27–41.
Vancouver: UBC Press.

McGurty, Eileen. 2007. *Transforming Environmentalism: Warren County, PCBs and
the Origins of Environmental Justice.* New Brunswick, NJ: Rutgers University
Press.

Murphy, Michelle. 2006. *Sick Building Syndrome and the Problem of Uncertainty:
Environmental Politics, Technoscience, and Women Workers.* Durham, NC: Duke
University Press.

Murphy, Michelle. 2017. "Alterlife and Decolonial Chemical Relations." *Cultural
Anthropology* 32 (4): 494–503.

Myers, Natasha. 2015. *Rendering Life Molecular: Models, Modelers, and Excitable
Matter.* Durham, NC: Duke University Press.

Native Youth Sexual Health Network and Women's Earth Alliance. 2016. *Violence
on the Land, Violence on Our Bodies: Building an Indigenous Response to Environmental Violence.* http://landbodydefense.org/uploads/files/VLVBReportTool
kit2016.pdf.

Oreskes, Naomi, and Erik M. Conway. 2010. *Merchants of Doubt: How a Handful of
Scientists Obscured the Truth on Issues from Tobacco Smoke to Global Warming.*
New York: Bloomsbury.

Plain, Wilson, Jr. 2016. *Aamjiwnaang First Nation Notification Report, May 2016.*
http://www.aamjiwnaang.ca/wp-content/uploads/2016/07/May-2016
-Notification-Report.pdf.

Proctor, Robert, and Londa Schiebinger, eds. 2008. *Agnotology: The Making and
Unmaking of Ignorance.* Stanford, CA: Stanford University Press.

Pulido, Laura. 2017. "Geographies of Race and Ethnicity II: Environmental Racism, Racial Capitalism and State-Sanctioned Violence." *Progress in Human Geography* 41 (4): 524–33. https://doi.org/10.1177/0309132516646495.

Simpson, Andre J., Buuan Lam, Miriam L. Diamond, D. James Donaldson, Brent A. Lefebvre, Arvin Q. Moser, Antony J. Williams, Nicolay I. Larin, and Mikhail P. Kvasha. 2006. "Assessing the Organic Composition of Urban Surface Films Using Nuclear Magnetic Resonance Spectroscopy." *Chemosphere* 63 (1): 142–52. https://doi.org/10.1016/j.chemosphere.2005.07.013.

Simpson, Audra. 2014. *Mohawk Interruptus: Political Life Across the Borders of Settler States.* Durham, NC: Duke University Press.

Spears, Ellen Griffith. 2016. *Baptized in PCBs: Race, Pollution, and Justice in an All American Town.* Chapel Hill: University of North Carolina Press.

Supran, Geoffrey, and Naomi Oreskes. 2017. "Assessing ExxonMobil's Climate Change Communications (1977–2014)." *Environmental Research Letters* 12 (8): 084019. https://doi.org/10.1088/1748–9326/aa815f.

Tuck, Eve. 2009. "Suspending Damage: A Letter to Communities." *Harvard Educational Review* 79 (3): 409–28.

Tuck, Eve, and K. Wayne Yang. 2013. "R-Words: Refusing Research." In *Humanizing Research: Decolonizing Qualitative Inquiry with Youth and Communities,* edited by Django Paris and Maisha T. Winn, 223–48. Los Angeles: SAGE.

United Church of Christ Commission on Racial Justice. 1987. *Toxic Wastes and Race in the United States: A National Report on the Racial and Socio-Economic Characteristics of Communities with Hazardous Waste Sites.* New York: United Church of Christ Commission on Racial Justice.

United Nations Environmental Programme. 2009. Annex: *Global Monitoring Report.* Conference of the Parties of the Stockholm Convention on Persistent Organic Pollutants, Geneva, May 4–8. http://chm.pops.int/Portals/0/Repository/COP4/UNEP-POPS-COP.4-33.English.PDF.

US Environmental Protection Agency. n.d. "Facts and Figures about the Great Lakes." Accessed September 18, 2015. https://www.epa.gov/greatlakes/great-lakes-facts-and-figures.

Wolfe, Patrick. 2006. "Settler Colonialism and the Elimination of the Native: Journal of Genocide Research." *Journal of Genocide Studies* 8 (4): 387–409.

Women's Earth Alliance and Native Youth Sexual Health Network. 2016. *Violence on the Land, Violence on Our Bodies: Building an Indigenous Response to Environmental Violence.* http://landbodydefense.org/uploads/files/VLVBReportToolkit2016.pdf.

PATRICK BRESNIHAN is Lecturer in the Department of Geography at Maynooth University. He works across the interdisciplinary fields of political ecology, science and technology studies, and environmental humanities. His research focuses on different but related concerns around water, energy, land, and infrastructure in Ireland. He has published extensively, including articles on community-water infrastructure in Cochabamba, urban commons in postcrash Dublin, and the poetics of John Clare. His book, *Transforming the Fisheries: Neoliberalism, Nature and the Commons* (2016), won the Geography Society of Ireland Book of the Year in 2018.

TIM CHOY is Associate Professor of Science and Technology Studies and Anthropology at the University of California, Davis. His writing and teaching have focused on the politics of atmospheres and experimental approaches to ethnography and concept work. His book *Ecologies of Comparison: An Ethnography of Endangerment in Hong Kong* was awarded the Rachel Carson Award by the Society for Social Studies of Science in 2013.

JOSEPH DUMIT is Chair of Performance Studies and Professor of Science and Technology Studies and of Anthropology at the University of California, Davis. He is an anthropologist of passions, brains, games, bodies, improvisation, drugs, and facts and has authored the books *Picturing Personhood: Brain Scans and Biomedical America*, and *Drugs for Life: How Pharmaceutical Companies Define Our Health*. He coedited three books and was the managing editor of *Culture, Medicine & Psychiatry* for ten years. His research and teaching constantly ask how exactly we come to think, do, and speak the way we do about ourselves and our world and what are the actual material ways in which we come to encounter facts, conspiracies, and things and take them to be relevant to our lives and our futures.

CORI HAYDEN is Associate Professor in the Department of Anthropology at the University of California, Berkeley, where she conducts research and teaches on the anthropology of science, technology, and medicine, with an emphasis on Latin America. Hayden is the author of *When Nature Goes Public* (2003) and is currently finishing a book, *The Spectacular Generic*, which analyzes the contours of a twenty-first-century politics of the pharmaceutical copy. She has also served as co-convener, with Tim Choy, Anne Walsh, and Julia Bryan-Wilson, of *Clouds and Crowds*, an initiative funded by the University of California Humanities Network on contemporary market formations, political atmospheres, aesthetic possibilities, and modes of inquiry.

STEFAN HELMREICH is Professor of Anthropology at the Massachusetts Institute of Technology. He is author of *Alien Ocean: Anthropological Voyages in Microbial Seas* (2009) and *Sounding the Limits of Life: Essays in the Anthropology of Biology and Beyond* (2016). His essays have appeared in *Critical Inquiry, Representations, American Anthropologist,* the *Wire, Cabinet,* and *Public Culture.*

JOSEPH MASCO is Professor of Anthropology and Science Studies at the University of Chicago. Most recently, he is the author of *The Theater of Operations: National Security Affect from the Cold War to the War on Terror* and *The Future of Fallout, and Other Episodes of Radioactive Worldmaking,* both from Duke University Press.

MICHELLE MURPHY is a Canada Research Chair in Science and Technology Studies and Environmental Data Justice, Director of the Technoscience Research Unit, and Professor of History and of Women and Gender Studies at the University of Toronto. Her current research concerns decolonial approaches to the study of chemical violence on the Great Lakes. She is Métis from Winnipeg and lives in Toronto.

NATASHA MYERS is Associate Professor in the Department of Anthropology at York University, director of the Plant Studies Collaboratory, and cofounder of Toronto's Technoscience Salon. Her current ethnographic projects speculate on protocols for seeding *Planthroposcenes,* with investigations spanning the arts and sciences of vegetal sensing and sentience, the politics of gardens, and the enduring colonial violence of restoration ecology. Her first book, *Rendering Life Molecular: Models, Modelers, and Excitable Matter* (Duke University Press, 2015) received the 2016 Merton Book Prize from the American Sociological Association's section on Science, Knowledge, and Technology. She is also the coauthor, with Carla Hustak, of the book *Le Ravissement de Darwin: Le langage des plantes* (2020).

DIMITRIS PAPADOPOULOS is Professor of Science, Technology, and Society and Director of the Institute for Science and Society at the University of Nottingham. He is currently a Leverhulme Fellow, and he is also the founding Director of EcoSocieties, one of the University of Nottingham's interdisciplinary research priority clusters. He is currently working on a monograph, *Chemicals, EcoPolitics, and and Reparative Justice*, and a coedited volume with María Puig de la Bellacasa and Maddalena Tacchetti titled *Ecological Reparation: Repair, Remediation and Resurgence in Social and Environmental Conflict* (2022). His most recent book is *Experimental Practice: Technoscience, Alterontologies, and More-Than-Social Movements* (Duke University Press, 2018).

MARÍA PUIG DE LA BELLACASA is Associate Professor at the Centre for Interdisciplinary Methodologies, University of Warwick, and an Arts and Humanities Research Council Leadership Fellow. Her most recent book, *Matters of Care: Speculative Ethics in More Than Human Worlds* (2017), initiates a conversation between feminist critical thinking on care and debates on more-than-human ontologies in science and technology studies, the environmental humanities, and social theory. Her current research concentrates on ongoing formations of ecological cultures, in particular around human-soil relations, looking at how connections between science, ecosocial movements, and artistic interventions contribute to transformative more-than-human ethics and politics. Based on this research, she is working on a new monograph, *When the Name for World Is Soil*.

ASTRID SCHRADER is Senior Lecturer in the Department of Sociology, Philosophy, and Anthropology at the University of Exeter, UK. She works at the intersections of feminist science studies, human-animal studies, new materialisms, and posthumanist theories. Schrader is particularly interested in scientific research on marine microbes, responsible knowledge production, time in the Anthropocene, interdisciplinarity, and relations between science and arts. Her work has been published in *Social Studies of Science*; *Environmental Philosophy*; *differences*; *Body and Society*; *Catalyst: Feminism, Theory, Technoscience*; and *Postmodern Culture*. She coedited a special issue of *differences* titled "Feminist Theory Out of Science."

ISABELLE STENGERS is Professor Emerita of the Université Libre de Bruxelles. Her first collaboration with Nobel Prize winner Ilya Prigogine led her to develop the contrast between the conceptual inventiveness of physics and its role as a model of objectivity. Claiming the irreducible plurality of scientific practices, she has proposed, as a challenge inseparably political and cultural, the concept of an active ecology of practices, embedded within a demanding, empowered environment. Her work is related to the philosophy of Deleuze, Whitehead, and James as well as to the anthropology of Latour and the SF thinking adventure of Haraway.

biochemical relays, 76–77

biodegradability, 26, 29, 52, 58, 165, 218

biofilm, 257

biogeochemical processes, 6, 12, 111, 201, 205, 214–15, 220; oceanic carbon cycle, 9, 121; soil and, 200, 209, 218; stories, 197, 198, 202, 203

biology, 7, 87, 93, 95, 120, 127n2, 215; ontology and, 110–11; of viruses, 115–16

biomolecular X-bonds (BXBs), 94–95

biopesticides, 40, 46

biopolitics, 108, 113, 114, 164, 170n1

bioremediation, 12, 29, 201, 205, 219–21, 224, 226n16

bios and *geos*, 109, 111, 112, 113, 120, 196

biosolids, 217

Black Lives Matter (BLM), 178, 248, 250, 253n5

Blackness, 260

Black Panther (2018), 43–44

Blencowe, Claire, 206–7, 225n6

Blumenberg, Hans, 96

bodies, 43, 102, 113, 201; Black, 187; degradation of, 214, 218; ecologies and, 45, 47–48; industrial chemicals in, 253n4, 271–73, 273; land and, 268, 271–72, 277; PCBs in, 258, 270; soil and, 208–10; spirit and, 204–5, 214; violence to, 271, 275, 276

Bogad, L. M., 249

Borch, Christian, 181, 185, 191n1, 192n2

Bord na Mona, 170n3

Boyle, Robert, 22

breakdown: as an ethos, 214, 216; assisting, 219, 221, 224; biogeochemical processes of, 6, 197, 198, 215; dictionary definition, 211; ecopoethics, 200–201, 220; embracing, 12, 211–13, 220; of food waste, 217–18; material-spiritual dimension, 215–16; mental health, 225n10; mottos, 207, 208; scale and, 226n11; of soil, 12, 125, 210–11, 213–14, 215

breathers: conspiracy and togetherness of, 13, 248–52, 251; drawings of, 242,

243–44, 245, 246–47, 247, 248; externality concept and, 239–40; lungs, 271, 275, 276, 277; as payers, 235–36, 240–42; respiratory illness, 242–43, 251

Bresnihan, Patrick, 3, 4, 10

Brighenti, Andrea, 179, 185

bromate, 100–101

brominated flame retardants (BFRs), 97–100

bromine (Br): bonds, 94; collagen scaffold, 86–89, *87*; description, 85; diatomic molecule, 27; main uses and toxicity of, 96–101; in pharmaceutical design, 94–95; researchers' methods, 8–9, 85–86

Brown, N. H., 89–90

Brown, Wendy, 179

Bubandt, Nils, 109, 113

building materials, 136, *137*

Butenandt, Adolf, 39

Canetti, Elias, 185–86, 189, 191

capitalism, 10, 11, 13, 48, 134; accumulation, 154, 191; colonial, 268, 271; investors, 157; petrochemical, 133, 135, 136, 146, 148; platform, 171n13; production, 114, 115, 125, 171n14; racial, 265, 269

carbon, 224n1, 240; cycle, 9, 109, 111, 114, 116, 121–24, 125; elemental, 7, 71, 75, 118–19, 196; imaginary, 10, 110–12, 114–15, 116, 118, 125–26

carbon dioxide (CO_2), 43, 121, 122–24, 203; emissions, 153, 157, 164

Carbon Neutral Laboratory for Sustainable Chemistry (Univ. of Nottingham), 43, 52

catastrophe, 11, 29, 31, 43, 221, 224

Cazdyn, Eric, 238

cells, 88–90, 120, 123, 272, 273

cesium, 138, *139*

chemical agents, 22–23, 24, 25

chemical models, 261, 272, 272–73, 275, 276

chemical practice: becoming elemental and, 61n14; composition and decomposition, 22–23; dose-dependent procedures, 28; ecology and, 8, 35–36, 37–38, 44, 45; eighteenth- and nineteenth-century, 22, 24–25, 32n2, 72; environmentally benign, 50; generative, 57–59; holistic, 51, 53; obliged, 49, 50; pharmaceutical design, 95; pollution and, 47, 48; scale and, 55–57

Chemical Valley (ON), 13, 258, 262–63, 266

Chemical Valley Emergency Coordinating Organization (CVECO), 263

chemistry, 1–2, 7, 73, 206, 215; causation and, 22–23; constitution of crowds and, 176, 183; ecological contingency of, 44, 53; environmental disasters and, 47; green and sustainable, 43, 49, 50–54, 58, 60n8; obligations of, 28–29; peace through, 53–54; physics and, 19–20; quantum, 24. *See also* periodic table

chemodiversity, 45

chemospheres, 36, 44, 47, 57

Chinese *wuxing* system, 2, 71

chlorine, 87

Choy, Tim, 2, 4, 13, 61n14

Clark, James, 52

Clark, Nigel, 113

classical elements, 2, 21–22, 25, 71, 80n2, 80n4, 202–4; crowds as, 186; love and strife and, 116–17; spirits of, 61n14. *See also* five elements

climate change, 6, 35, 124, 160, 203, 232; denial, 265; emissions targets, 157, 170n5; wind and, 10, 153, 164–65. *See also* global warming

CLORPT, 213

cloud (internet), 171n16, 188, 189

coal power, 159–60, 162, 171n17

Coase, Ronald, 239–40

Coca-Cola, 101

Cohen, Jeffrey Jerome, 76, 117–18

Cold War, 140, 142, 143

collagen, 86–89, 87, 90–92

collectivities, 236, 238, 272; conspiring, 237, 250, 252; political, 249

colonialism. *See* settler colonialism

color-blind ideologies, 266

commons, 4, 162, 172n20, 207; of breathing, 241; elemental, 12, 202; and uncommons, 242, 250, 251, 253n7

compost, 210, 211, 217–18

compounds, 25, 50, 53, 71, 95; artificial, 10, 136; breakdown of, 215, 217, 218, 219, 221, 224; chemical, 45, 216; healing, 57, 58; manufactured, 47, 221; of matter, 200, 201, 212, 213; nitrogen, 216, 226n14; organic, 44, 60n6, 98, 123; water, 203

concrete, 136, 137, 201, 212

Condillac, Etienne Bonnot, Abbé de, 72

connective tissue, 86–90, 91–92, 93–94

consent, politics of, 260, 273, 277

conspiracy, 2; of breathers, 13, 237, 247, 249–52

contagion. *See* emotional contagion

Cooper, Melinda, 172n21

cosmologies: classical elements, 2, 202–3; Earth origin, 196–98; Indigenous, 2–3

COVID-19, 248, 250–51

crowd theories: Canetti's, 185–86; Le Bon and Tarde's, 181–84; overview of, 11, 176–80, 191n1, 192n3; social media and, 188–91; waves and, 186–87

data infrastructures, 161–62, 266–67

DDT, 40, 59n2, 60n6, 77, 253n4, 258; Monsanto's manufacturing of, 269, 274

Dean, Jodi, 182–83, 185

death: by choke hold, 248, 251, 253n5; growth-reproduction and, 110, 111, 112; humanity's resistance to, 214; life and, 5, 6, 7, 152, 215; as regeneration, 208, 210

decolonialization, 4, 13, 59; approach to STS, 259, 268; of matter, 29, 165

decomposition, 122, 145, 216; degradation and, 214–15; of elements, 5–6, 7, 22–23, 26; of food waste, 217–18

deindividuation. *See* individuality

Deleuze, Gilles, 29, 61n13, 183, 192n4

democracy. *See* liberal/illiberal democracy

Derrida, Jacques, 125–26

Desert, the, 115

Diderot, Denis, 72

difference, 184, 192n4, 192n6, 250

digital media, 188, 192n8. *See also* social media

dissolved organic carbon (DOC), 122–24

DNA, 46, 119, 166, 168, 224n1

Don Quixote (Cervantes), 164

drinking water, 100–101

Duckert, Lowell, 76, 117–18

Dumit, Joseph, 3, 4, 8–9, 27, 61n12

Dupré, John, 120, 124

Durham, William F., 253n4

Durkheim, Émile, 8, 72–74, 78–79, 182, 184, 192n5

Earth, 5–6, 11, 28, 42–43, 61n14, 201; abandoning, 34–35; carbon and, 119; cosmologies, 196–97, 199, 202, 209; geological conditions, 134; plastics and, 144–45; saving, 222–23, 226n16

earth systems, 113, 121, 134, 136, 148; sciences, 142, 231–33

ecocommoning, 55, 211

ecocriticism, 9, 76

ecological degradation, 47, 54, 98, 142; plastics and, 145. *See also* pollution

ecological movements, 49, 55, 58, 60n7, 200

ecological sciences, 59n1

ecology, 5, 9, 12, 25–26, 28; chemical practice and, 8, 35–36, 39, 40, 45, 57–58; contingency and, 41–42, 44; environments and, 46; insurgent, 46–47; meanings and forms of, 37–38, 199–200; obligation and reparation and, 28, 36, 48–49, 58, 59, 220–21, 224, 252; political, 4, 169, 170n2; of practice, 61n11; scale and, 55–56; scientists and, 30–31; web of life, 27

economic theory, 239–40. *See also* environmental economics

ecopoethics/ecopoesis, 205, 224n2;

bioremediation, 220–21; breakdown, 200–201; concept, 198–99; license, 12, 202; soil, 200, 209, 214, 216–18

ecotoxicology, 28

electricity generation: grids, 158, 160–61, 171n10, 269, 274; in Ireland, 156–57; in Norway, 171n11; pylons, 154

elemental affinities, 12, 23, 198, 199, 221, 252; fostering, 200–201; material relations, 202–7; with soil, 208–11, 216–18

elemental thinking, 1–6, 9, 11; crowds and, 185, 187–88; eighteenth- and nineteenth-century, 70–74, 116–17, 180–81; feminist, 75; material relations and, 117–18

elements, definitions, 7, 18, 21–23, 72–73, 118. *See also* classical elements; five elements

emissions data, 266–67

emotional contagion, 180, 181, 184, 189–90, 192n3

Empedocles, 2, 21, 71, 116–17, 126, 202

Enbridge, 275, 276

endoscopy, 91–92

energy, 151, 170n3; coal power, 159, 160, 162, 171n17; narratives, 159–60, 165, 167, 169; smart grids, 155, 160–62, 171n13; storage, 161; wind, 153–54, 156–58, 162–65, 170n6

Eneropa region, 161

entailment, speculative, 238–39

environ, meaning, 76

environmental economics, 231–33, 235, 240, 241

environmental monitoring, 47, 261–62, 263, 266–67, 271

Environmental Protection Agency (EPA), 50, 100, 171n15, 267

epigenetics, 45–46, 270, 273

epistemologies, 22, 42, 44, 58, 178, 185, 261; colonial practices, 155; ecology, 37–38, 48; ontologies and, 2, 3

Erie, Lake, 258, 262

ethos/ethopoeisis, 39, 207, 211, 216

European Union (EU), 156–57, 160

existence, forms of, 110–11, 113, 115–16, 119, 126; Foucault on, 126n2

externality: from abstraction to aggregation, 237; breathers pay and, 235–36, 240–42, 249; conspiracy and, 252; economic theory and, 239–40; graphic representations, 231–34, *232, 234*

extinction, 32n3, 109, 115, 125

extracellular matrix (ECM), 89–90

extraction, 115, 163, 165, 265; Canadian banks and histories of, 274–75; labor, 211; logics of, 154–56, 170n2

Exxon-Mobil, 263, 265. *See also* Imperial Oil

Facebook, 188–90

Fanon, Frantz, 259–60, 265

farming, 38–39, 40–41, 70, 217–18

fascia, 90–93

fascism, 11, 177, 178

feedstocks, 51–52

feminist science studies, 3, 199, 216, 260

feminization, 182

fertilizers: nitrogen-based, 26, 42, 206, 216, 226n14; use of biosolids, 217–18. *See also* compost

finance capital, 274–75, 276

fire, 2, 11, 61n14, 71, 153, 203, 206; as a crowd, 186; Facebook as, 189; safety, 98–99. *See also* classical elements

five elements, 71, 80n4, 116–17, 203, 204

flame retardants, 97–100

floods, 165, 207, 225n6

Floyd, George, 248, 251, 253n5

fluorine, 85

food chain, 98, 116, 122, 123, 269

formaldehyde, 60n6, 71, 214, 253n5

Fortun, Kim, 242, 253n5, 254n9

fossil fuels, 60n6, 114, 153–54, 163, 206; extraction, 7, 115, 156, 275

Foucault, Michel, 110, 126n2

four elements. *See* classical elements

fracking fluid, 9, 100–101

freedoms, 179, 192n2

frontiers, 155, 160, 171n14

fruit flies: *Bactrocera oleae*, 38–39; *Drosophila*, 88

future(s), 126, 133, 251–52, 269–70; decolonial, 260, 268; Earth's, 5, 11, 144–45, 222; ecological, 200, 213; endurance and, 212; energy, 153, 160–61; imaginaries, 34, 43, 220; of life, 112, 125, 146

Galileo, 19, 31

Garner, Eric, 248, 251, 253n5

gaslighting, 13, 263–66, 268, 270, 276, 277

gasoline, leaded, 96

genes, 45–46, 120, 273. *See also* DNA

geology, 114, 136, 140, 144, 155, 215; periodization, 134, 139

geontology, 110, 111, 125

geontopower, 109, 116

geos, theories of, 108–10, 113

geosocialities, 108, 113–14

ghosts: of the Anthropocene, 108–9; Povinelli's collection of, 114–16; viruses as elemental, 9, 109, 118, 125–26

gifts, receiving, 30–31

global warming, 122, 126, 134, 163, 165

gnomes, 61n14

governance, 110, 124–25; colonial, 260, 262. *See also* sovereignty

Great Lakes, 13, 257–58, 269, 274

Greece, 39

green chemistry, 49, 50–54, 58, 60n8

greenhouse gases, 136, 156, 203

growth: biological process, 110, 111, 112; scale and, 54–55

Guardians of the Galaxy (2014), 43

Guattari, Félix, 61n13

Guimberteau, Jean-Claude, 91–93

Gut Feminism (Wilson), 102

Haber-Bosch process, 42, 43

Habtom, Sefanit, 250–51

halogens, 85, 97; bonds (X-bonds), 93–96

Han, Byung-Chul, 190

Haraway, Donna, 21, 74–75, 199, 211, 243, 247, 283

haunting, 109, 125–26

Hayden, Cori, 4, 11, 253n2

Hayward, Eva, 75

healing, 57–58, 220, 222–24

Helmreich, Stefan, 4, 8, 113, 124, 180, 186–87

Hirst, Paul Q., 73

Ho, P. Shing, 95

holistic practice, 51, 53, 91

Holland, 165, 172n18

Holocene, 134, 135, 142, 147

Hornsea wind project, 157–58

Howe, Cymene, 162, 164–65, 172n20

Hu, Tung-Hui, 191

Hudson, Billy, 89

Hudson Bay Company, 274

human/nonhuman relations, 42, 43, 59, 172n20, 212, 215; anthropogenic chemicals and, 36; land and, 268, 271–72, 277; soil and, 200, 208–11, 216, 217, 225n9; water and, 205–7, 258

Hunt, Nick, 152, 161

hydrogen bonds, 94, 96

hypnotism, 181–82

idiot, figure, 29–31, 164

imaginaries, 6, 164, 200; anthropocentric, 211; bioremediation, 220–21, 224; carbon, 10, 110–12, 114–15, 116, 118, 125–26; cosmic, 197; cultural, 4, 12; ecological, 209–10, 213, 220; elemental, 2, 61n14, 203–4, 219; future, 34, 43, 220; wind, 152–53

imitation-suggestion theories, 177, 181, 183–85, 189, 192n4

Imperial Oil, 262–63, 264, 265, 266, 275, 276

Indigenous peoples, 110, 115, 165, 204, 205, 250–51; cosmologies, 2–3; feminism, 260, 265; sovereignty, 2, 258, 259, 260, 268, 271, 273; stolen land, 13, 262–63, 266, 269

individuality, 23, 120; crowds and, 179–80, 181–85, 190; sovereign, 178, 191n2

industrial chemicals, 14; in bodies, 271–72, 273, 275, 276; damaged-based research, 267–68; decolonial approach to, 4, 258–61; as discrete entities, 261; emissions data and monitoring, 263, 266–67; gaslighting, 265; in water flows, 257–58. *See also* petrochemicals; polychlorinated biphenyls (PCBs)

industrialism, 6, 7, 12, 42, 146, 206; environmental impacts, 134; synthetic materials and, 136, 137

inflation, 185

infranatural forces, 214

infrastructures: chemical violence, 267, 271; ecochemical, 40–41, 59; energy, 7, 154, 162, 164; environmental data, 261–62, 275; of extraction, 274; gaslighting, 13, 265–66, 277

insecticides, 38, 60n2, 253n4; DDT, 40, 59n2, 60n6, 77, 253n4, 258, 269, 274. *See also* pesticides

instinctive attitude, 84

International Chronostratigraphic Chart (ICC), 147, 148

Interstellar (2014), 34–35

Ireland, 170n3, 171n12; magnitude of wind in, 10, 151–52; wind energy development, 153, 154, 156–58

Jackson, Mark, 216

Jackson, S. J., 213

James, William, 30

Jansen, Theo, 165–69, 166, 167, 172n18

Jue, Melody, 75

justice: environmental, 117, 212, 236, 241, 243–44, 266, 268; reparative, 8, 59; social, 35, 49, 101; worlding, 254n7

Kennedy, John F., 153

kinship relations, 74–75, 270, 275

Konsmo, Erin Marie, 271, 274

Kyoto Protocol, 156

Native Youth Sexual Health Network, 268, 271, 274

natural gas, 100

nature, elements of, 118, 141

naturecultures, 3, 6, 7, 14, 26, 199, 219, 223; achievements, 21, 24; elemental affinities and, 205, 207

Nausicaä of the Valley of the Wind (Miyazaki), 221–23

necrochemicals, 43

Neimanis, Astrida, 75, 205

Nelson, Diane, 235

neoliberalism, 131, 178–79, 190, 205, 212, 241

NIMBY politics, 154, 157

nitrogen, 203, 215, 224n1; fertilizers, 26, 42, 206, 216, 226n14; fixation, 42–43

nuclear age, 138–43

nuclear testing, 43, 60n3, 138–39, *139*, 142, 273

nutrient cycling, 121, 200, 209, 210, 215, 218

Oaxaca, Mexico, 164–65

obligation: to assist breakdown, 212, 217, 219; chemistry and, 28–29; collective, 43; ecological, 36, 48–49, 58, 59, 220–21, 224, 252; ecopoethical, 199, 200, 205, 216, 218; responsibility and, 6, 12; scientific practice and, 29–31

oceans and oceanography, 75–78; carbon cycle, 9, 111, 114, 116, 121–24, 127n5; life/nonlife in, 109; rising sea levels, 165; wind energy and, 151–52, 157–58, 163

O'Connor, Eddie, 170n3

oil refining, 262–63

olive farming, 40–41; fruit fly (*Bactrocera oleae*), 38–39

O'Malley, Maureen, 120, 124–25

Ontario, Lake, 13, 257–58, 274

ontologies, 9, 57, 61n3, 114, 118, 125, 199; alterontologies, 12, 36, 58; biology and, 110–11; ecological, 209; epistemologies and, 2, 3; ethos and, 39

Orban, Viktor, 192n2

origin stories, 196–97, 199, 200, 204, 208, 209

Orr, Jackie, 187

oxygen, 23, 27, 42, 97, 122

Palos Verdes Shelf (CA), 77

Papadopoulos, Dimitris, 8, 9, 25, 155, 163, 192n2, 200, 219; alterontologies, 5, 12, 254n7

Papadopoulos, Dionysis, 39, 40–41

Paracelsus, 61n14

Parker, Laura, 225n9

particles, 2, 60n6, 123, 226n11, 243; crowds as, 176, 177; microplastic, 145

particulate organic carbon (POC), 123

peace, 223, 224; and justice, 35; through chemistry, 53–54

Pentecost, Claire, 209–10

periodic table, 1–2, 8, 44, 59, 72, 118, 135; contemporary, 132; first sketch of, 70, 71; kinship relations and, 74; order of elements, 10, 23–24, 73–74, 132; plutonium of, 140; synthetic elements, 133

Perry, Alabama, 242

persistent organic pollutants (POPs), 13, 60n6, 97, 98, 257, 273. *See also* polychlorinated biphenyls (PCBs)

pesticides, 206, 269; synthetic and biopesticides, 40, 41, 46, 60n2, 60n6. *See also* insecticides

Peters, John Durham, 76, 188, 191, 192n8, 240

petrochemicals, 51–52, 144, 165, 169, 206; capitalism, 133, 135, 136, 146, 148; processing in Canada, 13, 258, 262–63, 266

pharmaceutical design, 94–95

pheromones, 38–41, 45

Phillips, Whitney, 189

photosynthesis, 111, 121, 122

physical reality, 19, 20

physicists, 19–20, 24, 27, 160

Pigou, Arthur Cecil, 239–40
pipelines, 263, 271, 275
planetary boundaries, 55, 60n10
plastics, 10, 26, 51, 56–57, 60n6; Barthes's essay on, 143–44; growth in production, 136, 137; prevalence and persistence of, 144–46; tubing, 165–67, 172n19; undecomposable, 216, 217–18
plastisphere, 145–46
Plato, 21
plutonium (Pu), 10, 43, 75, 136, 138–42, 146, 148
Poliakoff, Martyn, 52
pollution, 47, 48, 203, 205, 239; air, 4, 240, 242–44, 248, 254n9; batteries and, 171n15; under colonialism, 257, 259, 262; corporate perpetrators of, 274; data, 266–67; litter, 48; permission and, 13, 260, 262; plastics, 145
polybrominated diphenyl ethers (PBDEs), 97, 97–98, 99
polychlorinated biphenyls (PCBs), 13, 60n6, 77, 97, 97–98, 218; banning and persistence of, 269–70; in human bodies, 258; in Toronto's financial district, 257, 274
polyethylene (PE), 56, 144, 263; green, 51–52
polymers, 144, 145
Povinelli, Elizabeth, 109–11, 112, 114–16, 126
power formations, 114
property rights, 239
proteins, 86, 91, 93, 94–95, 119
Puig de la Bellacasa, María, 2, 5, 12, 31, 39, 49, 125, 252

racism, 11, 177, 178, 179, 182, 187; chemical relations and, 270; colonial, 260, 269; environmental, 266, 269. See also white supremacy
radioactive fallout, 138, 139, 244
rat model research, 242–44
reactivation, term usage, 1

rearrangement, elemental, 5–6, 7, 11–12, 14
reckoning. See atmospheric reckoning
Reclaiming tradition, 202, 204–5
recycling, 51–52, 56, 114, 209, 210; of food waste, 217–18; of organic matter, 112, 121, 123
reductionism, 72, 192n6, 206, 211, 212; environmental, 45–46
Reed, Aviva, 209–10
refractory dissolved organic carbon (RDOC), 124
reparation, 49, 57–59; ecological, 8, 36, 54, 59
reworlding processes, 31
Riordan, Rick, 80n4
Rules of Sociological Method, The (Durkheim), 72–73

Sagan, Carl, 224n1
Sagan, Dorion, 111
sailing ships, 158–59, 170n8, 171n9
Saint Clair River, 258, 262
salamanders, 61n14
Sarnia, Ontario, 262
scale, 61n12, 226n11, 253n6; growth and, 54–55; industrial, 42; intensive, 56–57; laboratory models and, 77–78; life/nonlife distinction and, 111; nested, 271, 276; planetary, 6, 133, 135, 138, 140, 146, 148, 263; renewable energy and, 163, 171n13
Schrader, Astrid, 3, 4, 9
Schwartz, M. A., 89
science and technology studies (STS), 3–4, 184, 199, 213, 242; decolonial approach to, 259, 268
science fiction, 39, 43–44, 48, 60n4, 221–23
scientific practice, 29–31, 224. See also chemical practice
Scribe, Megan, 250–51
Seaborg, Glenn, 140
semiochemicals, 41–42, 45
settler atmospherics, 250–51